Smart Agriculture

Volume 10

Series Editors

Zhao Zhang, Key Laboratory of Smart Agriculture System Integration Research, Ministry of Education, China Agricultural University, Beijing, China

Yiannis Ampatzidis, UF/IFAS Southwest Florida Research and Education Center, University of Florida, Immokalee, FL, USA

Paulo Flores, Department of Agricultural and Biosystems Engineering, North Dakota State University, Fargo, ND, USA

Yuanjie Wang, Agricultural Information Institute, Chinese Academy of Agricultural Sciences, Beijing, China

The book series Smart Agriculture presents progress of smart agricultural technologies, which includes, but not limited to, specialty crop harvest robotics, UAV technologies for row crops, innovative IoT applications in plant factories, and big data for optimizing production process. It includes both theoretical study and practical applications, with emphasis on systematic studies. AI technologies in agricultural productions will be emphasized, consisting of innovative algorithms and new application domains. Additionally, new crops are emerging, such as hemp in U.S., and covered as well. This book series would cover regions worldwide, such as U.S., Canada, China, Japan, Korea, and Brazil.

The book series Smart Agriculture aims to provide an academic platform for interdisciplinary researchers to provide their state-of-the-art technologies related to smart agriculture. Researchers of different academic backgrounds are encouraged to contribute to the book, such as agriculture engineers, breeders, horticulturist, agronomist, and plant pathologists. The series would target a very broad audience – all having a professional related to agriculture production. It also could be used as textbooks for graduate students.

Yande Liu · Liling Yang · Yinyan Shi ·
Guantian Wang · Dazhou Zhu · Zhao Zhang
Editors

New Technologies Applied in Apple Production

Sensing and Autonomous Systems

Springer

Editors
Yande Liu
Intelligent Mechanical and Electrical Equipment Innovation Research Institute
East China Jiaotong University
Nanchang, Jiangxi, China

Yinyan Shi
College of Engineering
Nanjing Agricultural University
Nanjing, Jiangsu, China

Dazhou Zhu
Institute of Food and Nutrition Development
Ministry of Agriculture and Rural Affairs
Beijing, China

Liling Yang
Institute of Agricultural Mechanization
Xinjiang Academy of Agricultural Sciences
Ürümqi, Xinjiang, China

Guantian Wang
School of Mechatronics and Vehicle Engineering
East China Jiaotong University
Nanchang, Jiangxi, China

Zhao Zhang
Research Center for Smart Agriculture
China Agricultural University
Beijing, China

ISSN 2731-3476 ISSN 2731-3484 (electronic)
Smart Agriculture
ISBN 978-981-97-7777-8 ISBN 978-981-97-7778-5 (eBook)
https://doi.org/10.1007/978-981-97-7778-5

This Springer imprint is published by the registered company Springer Nature Singapore Pte Ltd.
The registered company address is: 152 Beach Road, #21-01/04 Gateway East, Singapore 189721, Singapore

Preface

Apple is one of the most popular fruits, not only because of its good taste but also due to its benefits to health. Apple production has been heavily involved with labor in terms of harvest, pollination, and infield sorting. These manual operations significantly increase the apple production cost, as well as the retail price. A number of innovative sensing and automation technologies have been used in apple production to gradually replace the labor work. Specifically, artificial intelligence is transforming the conventional apple production approach. Though many papers, articles, theses, and dissertations have been published to describe new technologies applied in apple production, there lacks a book to comprehensively summarize the progress.

To fill the gap of missing a book to systematically provide new sensing and automation techniques applied in apple production, this book is proposed, which consists of seven chapters. The first chapter "Research Progress on Key Technology of Apple Harvesting Robots in Structured Orchards" is to discuss research progress on key technology of apple harvesting robots in structured orchards, followed by the second chapter "Design, Assembly and Test of a Low-Cost Vacuum Based Apple Harvesting Robot" on design, assembly and test of a low-cost vacuum based apple harvesting robot. In addition to harvest, pollination is another labor-intensive operation. Thus, the third chapter "Research Progress of Assisted Pollination Technology" focuses on research progress of assisted pollination technology, and forth chapter "Lightweight and Robust YOLOv5s with MobileNetV3 and GhostNet for Precision Apple Flower Detection for Pollination Drones" discusses lightweight and robust YOLOv5s with MobileNetV3 and GhostNet for precision apple flower detection for pollination drones. Then, the following two chapters "Apple Size Estimation Method with 3D Projection Correction for In-field Grading System" and "Research on the Fruit Movement Mechanism of Apple In-field Sorting Equipment" focus on apple infield sorting, with respective chapter titles of apple size estimation method with 3D projection correction for in-field grading system, and research on the fruit movement mechanism of apple in-field sorting equipment. In addition, apple quality detection is another important topic. The seventh chapter "Research on the Transferability of SSC Detection Models Between Different Instruments for Fresh Apricots" conducts research on the transferability of SSC detection models between different instruments for fresh apricots. Though it is on apricots, not directly related to apple, the research outcomes can be used as a reference for apple quality detection.

Drs. Yande Liu, Liling Yang, Yinyan Shi, Guanhua Wang, Dazhou Zhu and Zhao Zhang, the editors of this book titled *New Technologies Applied in Apple Production—Sensing and Autonomous Systems* have been working in the area of agriculture sensing and automation for decades, and their expertise and knowledge make a sound base for the quality of this book. This book provides state-of-the-art information for readers who are interested in learning about the important subject of innovative sensing and autonomous systems applied in apple production.

Dr. Yande Liu
Intelligent Mechanical and Electrical Equipment
Innovation Research Institute
East China Jiaotong University
Nanchang, China

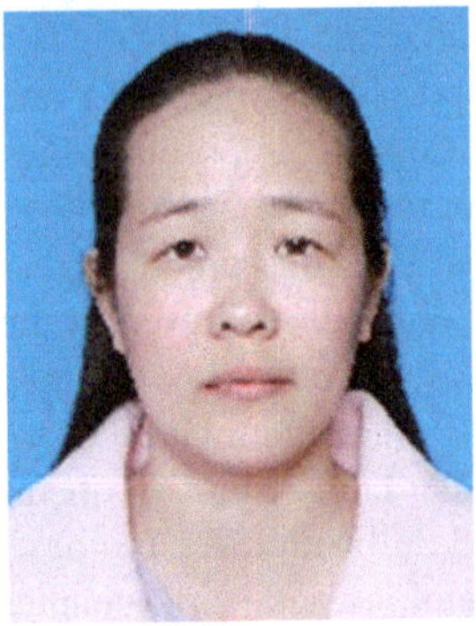

Dr. Liling Yang
Institute of Agricultural Mechanization
Xinjiang Academy of Agricultural Sciences
Ürümqi, China

Dr. Yinyan Shi
College of Engineering
Nanjing Agricultural University
Nanjing, China

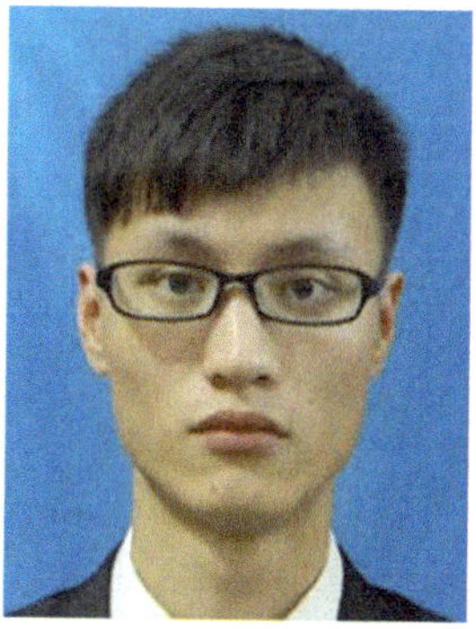

Dr. Guantian Wang
School of Mechatronics and Vehicle Engineering
East China Jiaotong University
Nanchang, China

Dr. Dazhou Zhu
Institute of Food and Nutrition Development
Ministry of Agriculture and Rural Affairs
Beijing, China

Zhao Zhang
Professor of College of Information and Electrical Engineering
China Agricultural University
Beijing, China

Contents

Research Progress on Key Technology of Apple Harvesting Robots in Structured Orchards

Wanjia Hua, Wenqiang Zhang, Zhao Zhang, Xiaohang Liu, Chayan Saha, N. S. Mustafa, and Dina Saber Salama

Abstract Currently, apple harvesting primarily relies on manual picking, which presents issues such as low efficiency, high costs, and frequent occupational health problems. As the labor shortage intensifies, the timely harvest of fresh apples faces significant challenges, making the use of robots to replace human labor an inevitable trend. With the groundbreaking advancements in standardized orchard production techniques and artificial intelligence technology, apple harvesting automation has achieved new breakthroughs. The key technologies of apple picking robots include apple recognition, apple positioning, robotic arms, motion planning, and end effectors. The performance of these technologies directly determines the efficiency and quality of the harvesting process. This paper provides an overview of the current development status of apple-picking robots. It reviews methods for apple recognition, introduces various sensor-based positioning technologies, studies robotic arms used for apple harvesting, discusses related motion planning algorithms, and analyzes end effectors and their detachment methods. The paper also introduces apple-picking robots that have already been commercialized and identifies the existing issues such

W. Hua · W. Zhang
China Agricultural University, Beijing 100083, China

Z. Zhang (✉) · X. Liu
Key Laboratory of Smart Agriculture System Integration, Ministry of Education, Beijing 100083, China
e-mail: zhaozhangcau@cau.edu.cn

Key Laboratory of Agricultural Information Acquisition Technology, Ministry of Agriculture and Rural Affairs, China Agricultural University, Beijing 100083, China

C. Saha
Department of Farm Power and Machinery, Bangladesh Agricultural University, Mymensingh 2202, Bangladesh

N. S. Mustafa
Pomology Department, Agricultural and Biological Research Institute, National Research Centre, Giza, Egypt

D. S. Salama
Water Relation and Field Irrigation Department, Agricultural and Biological Research Institute, National Research Centre, Giza, Egypt

Y. Liu et al. (eds.), *New Technologies Applied in Apple Production*, Smart Agriculture 10, https://doi.org/10.1007/978-981-97-7778-5_1

as long picking cycles, low success rates, severe fruit damage, and high costs. It points out future directions for commercial development, specifically focusing on reducing the complexity of orchard environments, optimizing recognition and motion planning algorithms, integrating multiple sensors, and enhancing multi-robotic arm cooperation.

Keywords Robotics · Apple picking · Research progress · Mechanization · Machine vision

1 Highlights

- Reviewed fresh market apple harvest technologies of robotics.
- Target recognition, localization, robotic arms, motion planning, end-effector are key technologies.
- Slow speed, low success rate, fruit damage, and high cost are key issues
- Optimization algorithms, sensors, and multi-arm coordination are future research directions

2 Introduction

With the increasing global demand, apple production has been steadily growing, reaching 83.1 million tons worldwide during the 2023 harvest season [68]. Given the long growth cycle of apples and the short maturity period of the fruit, the timing of harvesting is critical, necessitating a large seasonal labor force for the harvest. Efficient and timely apple harvesting is crucial for producing high-quality fruit and maximizing profits for growers. Harvesting costs now account for more than 25% of the total production costs [22], significantly reducing the economic returns for farmers. The apple industry is facing severe labor shortages and rising labor costs [45, 86], raising major concerns about its long-term sustainability and profitability. Consequently, mechanized harvesting technologies that replace manual labor are essential [57].

Mechanical apple harvesting technologies can be categorized into semi-automated harvesting, harvest-assist platforms, and robotic technologies [49, 83]. In semi-automated harvesting, the shake-and-catch method is used to harvest apples in batches [30, 31]. A significant challenge of this method is the high bruising rate of apples [51, 52]. Since the shake-and-catch method fails to resolve the bruising issue, researchers developed harvest-assist platforms to aid workers in apple harvesting, eventually leading to commercial applications [81, 82]. Although these platforms help increase worker efficiency, they still require manual picking [6]. Consequently, researchers have turned to developing harvesting robots to fully replace workers for the tedious, hazardous, and labor-intensive harvesting tasks [60, 71, 75, 87].

Since the 1970s, researchers have been developing robots for fruit harvesting [10, 28], primarily composed of vision sensors, robotic arms, end effectors, and mobile vehicles [55]. In 1985, France developed the first apple harvesting robot. However, progress in apple harvesting robots has been slow due to the unstructured field environments and complex interference conditions [76].

With the groundbreaking advancements in standardized orchard production techniques and artificial intelligence theories, apple production systems have undergone profound changes over the past two decades. Traditional unstructured orchards have been replaced by high-density orchards featuring smaller, structured trees, such as V-shaped trellises, vertical fruiting walls, and dwarf dense planting. These modern tree structures greatly facilitate orchard automation, leading to increased interest in exploring mechanized harvesting as a potential alternative to reduce the industry's reliance on manual labor and harvesting costs. Recently, researchers have conducted pioneering studies on autonomous intelligent apple-picking robots.

This paper first elucidates the current research status of apple-picking robots, summarizing the key technologies involved in their vision systems, robotic arms, motion planning, and end effectors. It highlights the existing problems and shortcomings of current apple-picking robots and offers prospects for future research directions.

3 Key Technology of Apple Picking Robot

Apple-picking robots use vision systems to identify and locate apples, and end effectors to detach the fruit. The challenge in developing apple-picking robots lies in accurately identifying and locating the fruit, and detaching it from the tree without causing damage to either the fruit or the tree.

3.1 Current Status of Apple Recognition Research

The primary step in apple picking is the identification of the apples. Apples grow in various orientations in their natural state, and due to the angles of the harvesting robots, the images captured of the apples can be diverse. These images may feature unobstructed views, leaf obstructions, or overlapping fruits, all of which can affect the accuracy of the apple-picking robot's vision system in identifying the fruit. The commonly used target recognition algorithms include traditional image processing techniques, machine learning-based image processing techniques, and deep learning technologies.

3.1.1 Traditional Image Processing

Traditional image processing methods rely on features such as color, shape, and texture of the image. These methods employ algorithms like threshold segmentation and color difference to segment the image. In these approaches, predefined rules and specific image processing techniques such as edge detection and morphological operations are used to process and analyze the image, extracting the contours and features of the target object.

Bulanon et al. [7] utilized color difference and optimal thresholding for segmentation and recognition of Fuji apples. The color difference measurement in this method aimed to enhance the image, making the optimal threshold dependent on the maximum grayscale variance of the color difference. Experimental results showed an overall recognition success rate exceeding 80%, but color-based recognition failed to eliminate the effects of changing lighting conditions. Specifically, under backlight conditions, the error rate of the recognition algorithm reached 18%. Wachs et al. [72] employed maximum mutual information from infrared and color images to extract high and low-level visual features, thereby detecting "green" apples in the foreground of fruit tree canopies. The recognition rates of the two features reached 54% and 74%, respectively. Jiang and Zhao [38] utilized models of two different colors to estimate the yield of mature apples. The R2 values between the predicted apple quantities by the models and the actual harvest quantities ranged from 0.58 to 0.71. However, the accuracy of the system was affected by background object occlusion. Nguyen et al. [50] utilized color and shape attributes to identify apples in point clouds obtained from Kinect v1. The results showed that 100% of fully visible apples and 82% of partially occluded apples were correctly detected. Feng et al. [18] employed a forward-looking infrared (FLIR) camera based on multispectral dynamic imaging (MSX) technology to capture apple images. They proposed an effective potential apple region locking algorithm based on the pseudo-color and texture information of MSX images.

Researchers have conducted numerous studies on factors such as changes in lighting conditions, overlapping occlusions, and apples of the same color. They have manually crafted features such as color, texture, and shape to identify apples in specific scenarios. This approach can effectively recognize and segment targets in simple scenes to some extent. However, for complex backgrounds or varied fruit conditions, its recognition accuracy and robustness are poor, making it challenging to meet the demands of practical applications.

3.1.2 Machine Learning Based Image Processing

Machine learning techniques involve key steps such as feature selection, preparing training data, and model selection. By employing algorithms like K-means clustering, region growing, and support vector machines, machine learning techniques can utilize extensive training data to automatically adjust model parameters, thereby enhancing recognition performance.

Tabb et al. [62] developed the Global Gaussian Mixture (GMOG) algorithm to process real-time video sequences captured by a row-based harvesting machine. The algorithm executes at a speed of 14–16 frames per second and correctly identifies 85–96% of red and yellow apples. Ji et al. [35] proposed a segmentation method based on the growth method of seed areas and color features to process two-dimensional images of apple trees. They also employed a support vector machine classifier for apple identification, achieving a success rate of 89%. Ma et al. [47] introduced an algorithm that combines the global search capability of the quantum genetic algorithm with the adaptability of fuzzy reasoning neural networks to identify apple fruits. The correct recognition rate for overlapping images is 92.31%. Kelman and Linker [] determined the edges of apples through convexity detection of tree images and performed three-dimensional modeling using a least squares constraint mechanism. This method achieved a localization accuracy of 94% for overlapping apples but had poor real-time performance, with an average processing time of approximately 2 min per image. Wang et al. [69] addressed the issue of existing segmentation methods for overlapping apples not preserving contour information of the overlap. They comprehensively used K-means clustering algorithm, Ncut algorithm, and Spline interpolation algorithm for segmentation of overlapping apple targets, achieving average segmentation error and average overlap of 5.24% and 93.81%, respectively. Lv et al. [46] proposed an apple image segmentation algorithm based on convex hull center and Markov. They employed the Otsu algorithm to obtain thresholds and performed adaptive segmentation of significant regions for apple images.

Machine learning-based image processing techniques can better adapt to target recognition tasks in complex scenarios and possess a certain degree of generalization ability. This enhances the robustness and reliability of the visual systems of apple-picking robots, enabling them to handle slightly overlapping or shaded fruits effectively. However, for the issue of large overlapping areas of fruits, there is currently no optimal solution. Further research is needed to improve the recognition rate and localization accuracy of heavily overlapping fruits.

3.1.3 Deep Learning

To further enhance the performance of apple-picking robot visual systems, deep learning technology has emerged as a crucial development direction. Deep learning techniques leverage deeper neural network architectures to automatically learn and extract richer and higher-level features, thereby improving the accuracy and robustness of image processing. Particularly in addressing complex scenarios such as overlapping fruits, deep learning models excel in understanding and interpreting intricate features in images, enabling precise recognition and localization of overlapping fruits.

In the realm of apple target detection for picking robots, Sa et al. [54] proposed a novel multimodal Faster R-CNN model, combining RGB and NIR information for fruit detection tasks, achieving an accuracy of 83.8% across 7 fruit types. Suchet and James [59] applied the Faster R-CNN object detection framework to apple detection, leveraging transfer learning and PCA data augmentation techniques. With minimal dataset requirements, they achieved a detection success rate of 90.7%. Tian et al. [65] introduced an improved YOLOv3 network based on the DenseNet method to detect apples at three different growth stages in orchards, effectively detecting apples under overlapping and occluded conditions. Kang and Chen [89] proposed an apple detection model based on the LedNet network, enhancing detection performance with Feature Pyramid Networks (FPN) and Atrous Spatial Pyramid Pooling, achieving real-time fruit detection. Gené-Mola et al. [27] utilized instance segmentation neural networks and Structure from Motion (SfM) for fruit detection. They projected MASK R-CNN detections onto a 3D point cloud based on SfM, improving system performance from an F1 score of 0.816 (2D fruit detection) to 0.881 (3D fruit detection and localization). Gao et al. [23] employed Faster R-CNN to detect ripe apples under various occlusion conditions, achieving an average precision of 87.9% in non-occluded, leaf-occluded, branch-occluded, and inter-fruit occluded scenarios, with a detection time of 24.1 ms. Jia et al. [36, 37] presented an improved Mask R-CNN-based visual detection model for harvesting robots, achieving an accuracy of 97.31% and a recall rate of 95.70%. Chu et al. [13] developed an apple detection framework based on Mask R-CNN for detecting apples during the harvesting season, achieving an F1 score of 0.905 with a detection time of 0.25 s per frame. Chen et al. [12] devised an apple detection method based on Des-YOLOv4, utilizing Soft-NMS to address missing detection issues and improve detection of missed fruits. Lu et al. [44] proposed an improved YOLOv4-based apple detection model, enhancing detection accuracy by incorporating convolutional attention modules, achieving detection of both immature and ripe apples. Praveen and Naveen [] introduced a new approach utilizing drone images and YOLOv7 for apple detection. With the introduction of multi-head attention mechanism, the precision of the object detection model was significantly improved, enabling more accurate estimation of apple depth.

Deep learning faces challenges in apple recognition as well. Especially when dealing with apple images captured in natural environments, recognition performance remains suboptimal for apples with unique growth postures such as variations in lighting, occlusions, and overlapping. Particularly, identifying and segmenting apple edges poses significant difficulty. Additionally, while improving algorithmic precision in recognition and segmentation, it leads to decreased real-time performance, thus affecting the efficiency of fruit-picking robots, failing to meet the predefined requirements.

3.2 Apple 3D Positioning and Its Sensors

After acquiring apple targets from the image recognition system, apple-picking robots need to obtain the three-dimensional positions of the apple fruits. Common depth cameras for obtaining depth information include stereo cameras, structured light cameras, and time-of-flight (ToF) cameras.

3.2.1 Stereoscopic Camera

Based on the principle of stereo matching, using two or more cameras to capture images from different angles recreates a three-dimensional scene. The principle of stereo vision 3D matching is depicted in Fig. 1. By analyzing the differences in images from different viewpoints, depth information can be obtained. Depth information can be easily acquired at low cost and high resolution.

Takahashi et al. [63] used stereo vision technology to identify apples in orchards. First, they divided the three-dimensional space into several cross-sections based on the line-of-sight distance. The advantages of stereo or multi-view stereo vision matching lie in high image resolution and precision in measuring objects at close distances. However, drawbacks include susceptibility to changes in lighting conditions and object textures, making it unable to function properly in dark environments. Additionally, the stereo matching and calibration processes are relatively complex, and the computation time is long. Then they reconstructed the three-dimensional information of objects by integrating the two-dimensional images of these cross-sections. Experimental results demonstrated that using this method for depth measurement, the error was controlled within the range of ± 5%, with a fruit identification rate greater than or equal to 90%. Li et al. [41, 42] employed a stereo vision system to obtain the 3D positions of apples. This method initially employs Faster R-CNN for apple detection from the images captured by the visual system. Then, based on color features, it segments apples from the images, followed by stereo matching to identify the three-dimensional spatial information of apples. However, this method performs poorly when fruits are occluded.

The advantages of stereo or multi-view stereo vision matching lie in high image resolution and precision in measuring objects at close distances. However, drawbacks include susceptibility to changes in lighting conditions and object textures, making it unable to function properly in dark environments. Additionally, the stereo matching and calibration processes are relatively complex, and the computation time is long.

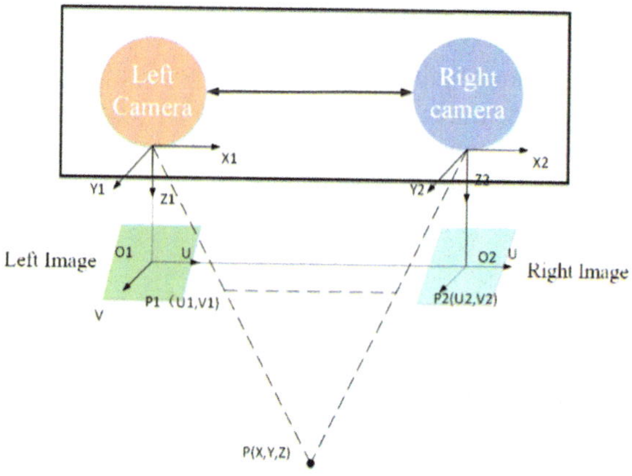

Fig. 1 Stereo vision schematic

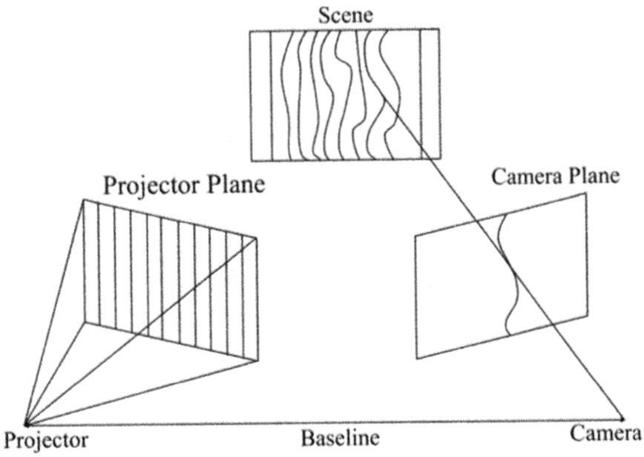

Fig. 2 Structured light camera schematic

3.2.2 Structured Light Camera

Figure 2 illustrates the fundamental principle of a structured light camera for acquiring depth information [2]. A structured light camera typically comprises one or more monocular cameras and a projector. The projector emits grating patterns onto the surface of the object under measurement. Subsequently, upon receiving a trigger signal, the binocular camera captures the reflected patterned image [3].

Andriyanov et al. [1] proposed an apple spatial position intelligent estimation system based on YOLOv3 and RealSense D415 camera. This method can acquire high-precision apple position information, with an average mean square error ranging from 7 to 12 mm within a measurement range of 1.5 m, and an average positioning speed of 2.5 frames per second. Due to the significant error inherent in directly using the centroid of the fruit surface point cloud instead of the fruit centroid, Gao et al. [25] utilized the RANSAC algorithm to fit the surface point cloud of the fruit obtained by the FM811-IX-A camera with a sphere, thereby obtaining the centroid of the fruit and enhancing the accuracy of fruit positioning. The standard deviation of the fitted sphere radius for apples was 1.79 mm, and the standard deviation of the sphere center coordinates was less than 2.5 mm.

Structured light cameras can rapidly acquire depth information and are not affected by image texture features. However, their effective measurement distance is limited, and as the distance increases, the projected pattern becomes larger, leading to reduced accuracy. They are not suitable for outdoor use in strong sunlight, as intense light can interfere with the projection.

3.2.3 Time of Flight (ToF) Camera

The ToF camera obtains distance information by measuring the time it takes for light to travel from the camera to the target and back. Specifically, it continuously emits laser pulses onto the target and receives the reflected light using sensors. By measuring the round-trip time of the light pulse, which is the time it takes for the laser pulse to be emitted from the camera and return to it, the depth of the target is calculated. Since the speed of light is extremely fast, directly measuring the flight time of light is impractical. Therefore, ToF cameras typically employ a modulation technique to modulate the light pulses, and then measure the phase shift of the modulated light wave to calculate the flight time, thereby obtaining depth information about the target.

Song [58] proposed a method to determine the three-dimensional coordinates of fruits for guiding robotic arms in harvesting, based on the centroid coordinates obtained from a laser scanner. Experimental results showed that within a range of 250–650 mm from the target, the average positioning time was 0.34 s, with an average positioning error of 11.2 mm. Nguyen et al. [50] utilized a Kinect sensor for apple detection based on color features, achieving a detection success rate of 100% for unobstructed apples and 80% for partially obstructed ones. The localization error was less than 1.0 cm, with a single fruit detection time of 50 ms. Tu et al. [67] employed a Kinect v2 camera combined with the MS-FRCNN convolutional neural network to study the identification and localization of fruits under variable lighting conditions and occlusion. Gené-Mola et al. [26] introduced a novel method utilizing a Mobile Terrestrial Laser Scanner (MTLS) with reflective capability, for detecting and positioning Fuji apples in real commercial orchard environments. Tsoulias et al. [66] utilized LiDAR to locate fruits based on backscatter intensity and geometric features, while estimating their sizes. Sun et al. [61] improved the density peak clustering segmentation algorithm for target fruit localization and identification using the gradient field of depth images. This method was designed to be non-iterative, non-hierarchical, and single-sample, significantly enhancing operational efficiency.

Even when the differences in texture features are weak, depth information can be quickly obtained. However, its accuracy is relatively poor and it is difficult to achieve millimeter-level precision, which is not as good as the near-distance (within 1 m) application of structured light. Its applications are limited to the field of high-precision at close distances, and it can hardly be used in strong outdoor lighting conditions.

3.2.4 Fusion of Multiple Sensors

With the rapid growth in demand for robot systems in agricultural settings, there is an urgent need for depth sensing devices capable of accurate and robust fruit localization, along with corresponding perception methods. In this scenario, information fusion from different cameras holds great promise for achieving better perception capabilities.

Si et al. [56] were among the first to explore fruit localization methods, using a stereo vision system and a PD4 laser rangefinder to achieve three-dimensional spatial localization of apples. Kang et al. [39] proposed a visual perception method based on high-resolution LiDAR-camera fusion, achieving fruit localization standard deviations of 0.253, 0.230, and 0.285 cm at distances of 0.5, 1.2, and 1.8 m, respectively, even under strong sunlight conditions in the afternoon. Zhang et al. [80] developed an Active Laser Camera Scanner (ALACS) system consisting of a red line laser, an RGB camera, and a linear motion slider, enabling high-precision localization with a maximum depth measurement error of less than 4 mm at distances ranging from 0.6 to 1.2 m.

3.3 Current Status of Research on Robotic Arms for Harvesting Robots

Once the vision system completes apple recognition and localization, it transfers the position information to the robotic arm, which then moves the end effector to the picking position. The robotic arm structures applied to apple picking mainly consist of three types: serial joint robotic arms, parallel robotic arms, and Cartesian robotic arms (Fig. 3).

The tandem arm technology is mature and flexible, widely used in agricultural settings. Currently, most harvesting robots employ articulated arms as their main body [71, 90]. However, their working space is fixed and may not align well with the distribution of fruit on apple trees. Researchers have independently designed a series of arms [33, 57, 84].

Parallel manipulator arms possess strong load-bearing capabilities, high precision, and minimal end-effector inertia, ensuring both harvesting efficiency and accuracy. They are particularly suitable as the main body for vacuum harvesting robots. However, their range is limited, preventing them from penetrating the canopy.

Cartesian manipulator arms move linearly along the three coordinate axes of a Cartesian coordinate system, allowing for high positioning accuracy, a large workspace, and low manufacturing costs [41, 42, 73]. However, they have poor obstacle avoidance capabilities.

Currently, most harvesting robots in China opt for industrial manipulator arms, offering advantages such as shortened development time, long service life, high positioning accuracy, complete support programs allowing for reprocessing on existing programs, and effective improvement of harvesting speed due to well-coordinated rotational inertia. However, this system also has drawbacks, including high costs, high power consumption, and inability to adapt to complex and changing harvesting

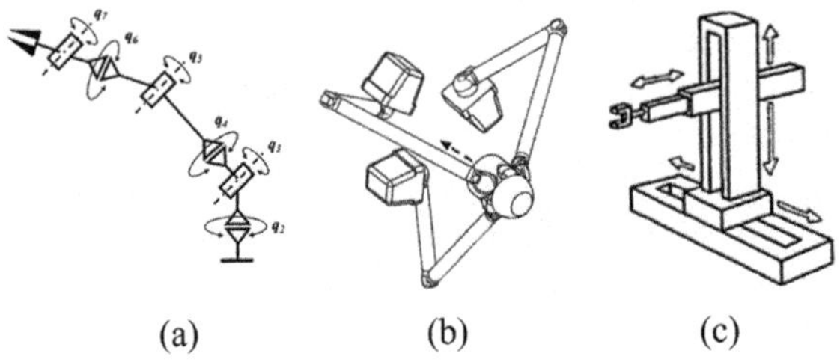

Fig. 3 Apple harvesting robot arm: **a** tandem robotic arm, **b** parallel robotic arm and **c** Cartesian robotic arm

environments. In contrast, self-designed manipulator arms based on industrial robot structures suffer from several problems, including high power consumption, high cost and poor adaptability to the working environment. However, they may extend the development cycle and have lower efficiency compared to industrial manipulator arms.

3.4 State of the Art in Robotic Arm Motion Planning Research

Motion planning for robots aims to construct a motion trajectory from an initial position to a target position, typically divided into path planning and trajectory planning. Motion planning consists of path planning (in space) and trajectory planning (in time), where the sequence of points or curves connecting the start and end positions is referred to as the path, and the strategy that constructs the path is termed path planning. For fruit and vegetable harvesting robots, path planning involves finding an obstacle-free path. Common intelligent obstacle avoidance algorithms include RRT, A* algorithm, artificial potential field method, ant colony algorithm, deep reinforcement learning, among others. Trajectory planning ensures the stability of the manipulator while completing tasks and is mainly applied using polynomial functions, B-spline functions, etc. Path planning precedes trajectory planning and serves as the foundation for trajectory planning.

3.4.1 Harvesting Robot Path Planning

For apple harvesting robots, the efficiency of harvesting mainly depends on the path planning for picking. The complex environment of apple fruit harvesting makes optimizing robot harvesting performance through obstacle avoidance path planning an urgent problem to be solved. This section introduces five mainstream path planning algorithms and their recent improvements.

A. Sample-Based Path Planning

The Rapidly-exploring Random Tree (RRT) algorithm based on node sampling achieves effective path planning for high-degree-of-freedom, complex-structure manipulators. The principle of the RRT algorithm is to establish a path search tree at the starting point and search random points by random sampling. If there is no collision, the set step length is extended toward the direction of the random point until the goal point is added to the extended tree, thereby determining a collision-free path. The RRT algorithm has high optimization efficiency in multidimensional space, but it has strong randomness, low node utilization rate, unstable paths, and cannot guarantee that the found path is the optimal path. In response to the shortcomings of the RRT algorithm, researchers have made some progress in improving the algorithm in recent years. Cao et al. [9] proposed a genetic algorithm (GA) and smoothing method to optimize the path generated by RRT. The path length was shortened by 20%, and no collisions occurred in real-world experiments. Chen et al. [11] proposed a feasible region local search path planning method for fruit-picking robots based on discrete workspace guidance and configuration space exploration correction. The algorithm can achieve a practical picking rate of 10.5 items/second, significantly improving the path quality in dynamic and complex agricultural scenes.

Path planning for a fruit-picking robot involves determining an obstacle-free route from the starting point to the fruit location. Path planning methods play a crucial role in enhancing the robot's performance. This section outlines five mainstream path planning algorithms and their recent advancements. In apple harvesting robotics, the efficiency of harvesting is primarily contingent upon the picking path planning. The intricate environment associated with apple fruit harvesting underscores the pressing need to optimize robot harvesting performance through obstacle avoidance path planning.

B. Graph-Based Path Planning

Graph-based path planning methods first require building an environment map and then constructing the optimal path from the starting point to the destination based on the map information. After knowing the environment information, the A* algorithm is commonly used to find the optimal path. The A* algorithm combines the Dijkstra algorithm and the greedy search algorithm, determining the selection of child nodes based on the weighted sum of the actual cost from the start point to the current point and the estimated cost from the end point to the current point. The A* algorithm is often combined with grid methods. Although guided by the heuristic function, A* algorithm can reach the goal faster, this method tends to result in excessive turning, making the planned path convoluted. Additionally, the computation speed of the A* algorithm is slow, which affects the picking efficiency in practical applications. Tang et al. [64] proposed an obstacle avoidance algorithm for 6-DOF robotic arm based on improved A* algorithm and artificial potential field method. Finally, path avoidance tests were conducted using the AUBO-i10 robotic arm. The experimental results demonstrate that the proposed method can make the manipulator move smoothly, effectively plan obstacle-free paths, and prove the feasibility of the method.

C. Artificial Potential Field-Based Path Planning

The artificial potential field method was proposed by Khatib in 1986, creating virtual attractive potential fields around target points and virtual repulsive potential fields around obstacles. The combined effect of the two potential fields forms a new potential field, namely, the artificial potential field. The artificial potential field method has advantages such as low computational complexity, simple structure, and ease of control. However, when the resultant force is zero, the robot will be trapped in a local optimal state, and when obstacles are close to the target point, the target may become unreachable. Traditional artificial potential field methods are prone to failure in planning due to the complex obstacles in the working environment of fruit and vegetable harvesting robots. Xie et al. [74] transferred the application principle of two-dimensional potential fields to three-dimensional space, obtaining smoother obstacle avoidance motion curves. Zhuang et al. [88] proposed an obstacle avoidance path planning method for apple harvesting robot arms based on APF-A. This research method is highly feasible and effective in planning actual obstacle avoidance paths for apple picking robot arms, effectively reducing the length of obstacle avoidance paths and planning operation time.

D. Ant Colony Algorithm-Based Path Planning

The Ant Colony Optimization (ACO) algorithm is an intelligent algorithm that simulates ants' selection of walking directions based on the concentration of pheromones when searching for food. Due to the lack of initial pheromones and the indistinct differences in heuristic information, traditional ant colony algorithms require long search times, and when the grid model is complex, they are prone to deadlock. The resultant paths planned by traditional ant colony algorithms often have many bends and large cumulative bend angles. Researchers have improved the ant colony algorithm or combined it with other algorithms to obtain better paths. Zhang et al. [77, 79, 84] proposed the Optimal Sequence Ant Colony Optimization algorithm (OSACO) for the path planning problem of continuous fruit picking, which can adjust parameter combinations intelligently at different scales, thus giving full play to the optimal performance of the ant colony algorithm. The average path length of the OSACO algorithm was reduced by 6.18%, demonstrating better global search capability, higher-quality convergence to the optimal solution, shorter generated path length, and stronger robustness.

E. Deep Reinforcement Learning-Based Path Planning

The Deep Reinforcement Learning (DRL) algorithm performs well when facing complex environments and tasks, aiding in autonomous learning and obstacle avoidance planning for robots. This algorithm utilizes the end-to-end mapping capability of Deep Learning (DL) to obtain strong perception and optimizes strategies through Reinforcement Learning (RL) by self-learning from interactions with the environment to obtain feedback, possessing powerful perception capabilities of deep learning and intelligent decision-making capabilities of reinforcement learning. Li et al. [43] proposed a collision-free stalk grabbing method for tomato harvesting robots based on deep reinforcement learning. In field experiments conducted in tomato greenhouses, the robot's harvesting success rate was 85.5%. Compared with traditional methods with fixed horizontal and parallel to the main stem postures, the success rates increased by 57.3% and 43.0%, respectively. The average operation time from identification to successful harvesting was 11.42 s, providing a promising solution to improve the efficiency of tomato harvesting robots.

3.4.2 Harvesting Robot Trajectory Planning

Improper harvesting trajectory planning can lead to unstable robot arm speeds and accelerations, increasing the failure rate of picking. Therefore, on the basis of path planning, trajectory calibration of the manipulator is necessary to achieve efficient and stable picking.

Trajectory planning is divided into Cartesian space trajectory planning and joint space trajectory planning [24]. Cartesian space trajectory planning refers to planning motion trajectories in Cartesian coordinate systems, allowing robots or other motion systems to execute motions between given start and target points. In Cartesian coordinate systems, positions are represented by coordinates (x, y, z), and trajectory planning involves determining the variation of these coordinates over time. The object of Cartesian space trajectory planning is generally the motion quantities of the end effector of the harvesting robot, such as displacement, velocity, acceleration, etc., which can intuitively reflect the motion of the end effector of the robot. Cartesian space trajectory planning is suitable for point-to-point trajectory planning, including linear planning [48], circular arc planning [70], polynomial curve planning [5], spline curve planning [29], and "S" velocity curve planning [20].

Joint space trajectory planning refers to planning motion trajectories in the joint space of robots or motion systems, determining the variation of joint angles over time to achieve the required motion tasks. Compared to Cartesian space trajectory planning, joint space trajectory planning is more direct and can more accurately control the robot's motion. Its objects are the robot's joint variables, joint angular velocities, joint angular accelerations, and other motion quantities. However, joint space trajectory planning cannot intuitively reflect the motion trajectory of the robot's end effector and does not impose constraints on its motion trajectory. Joint-type trajectory planning methods are generally used more in continuous motion scenarios without specific path requirements, including polynomial interpolation [53], trapezoidal velocity interpolation [21], spline curve interpolation, etc. [34].

Cartesian space trajectory planning constructs displacement, velocity, and acceleration curves of the end effector of the manipulator over time based on its actual motion. Joint space trajectory planning uses fitting functions to construct trajectory curves that vary over time by constraining joint variables, velocities, accelerations, and other combined variables. Regardless of the type of space trajectory planning, it is essential to ensure that the first two derivatives of robot motion are continuous.

3.5 State of Research on End-Effectors for Apple Separation

The end effector is a critical component of apple harvesting robots. After apples are identified, located, and their harvesting routes determined, the end effector is used to separate the fruit from the branches. Due to the fragile nature of apple skin and the complexity of the growing environment, apples are susceptible to damage during harvesting. The flexibility of the end effector, its method of securing the fruit, and the way it separates the fruit are crucial factors influencing the extent of fruit damage. Therefore, developing end effectors that reduce apple damage rates is essential. There are many types of end effectors for agricultural robot harvesting, which are currently in active development. Given the complexity of apple growth

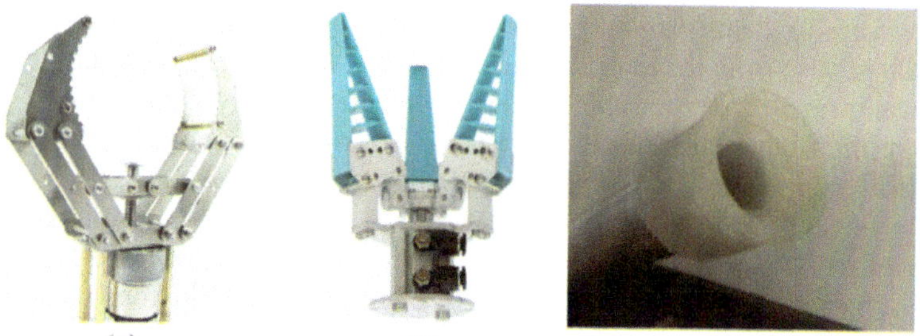

Fig. 4 End-effectors for apple separation: **a** bionic rigid clamping type, **b** bionic flexible clamping type and **c** negative pressure adsorption type

environments and the difficulty in locating and positioning fruit stems, commonly used end effectors for separating fruit include: biomimetic mechanical rigid grippers, biomimetic pneumatic flexible grippers, and vacuum suction types (Fig. 4). Each type of end effector has its advantages and disadvantages.

Biomimetic mechanical rigid grippers mimic human picking actions and separate fruit through twisting and pulling motions. They have low positioning requirements for apples and are suitable for apples of various sizes. However, due to branch obstruction, there is a risk of simultaneously grasping branches and fruit during fruit retrieval, leading to failed apple picking or even damage to the gripper. Biomimetic pneumatic flexible grippers exhibit a robust gripping force and are capable of maintaining a stable grasp on a diverse range of fruit, irrespective of their size and shape. However, they require air sources and pneumatic control systems, adding complexity to the system. Precise control and adjustment are needed to avoid fruit damage or unstable gripping.

Baeten et al. [4] designed a novel end effector for apple harvesting, consisting mainly of a vacuum suction cup and an internal camera. It utilizes vacuum suction to harvest apples, with harvesting experiments showing a success rate of 80% and an average harvesting time of 8–10 s. However, it resulted in severe fruit damage and exhibited poor adaptability to fruits of different sizes, lacking stability. Silwal et al. [57] developed an underactuated apple harvesting end effector designed specifically for modernized orchard structures. The end effector utilizes a tendon-pulley transmission system, comprising three drivers and three fingers, each with two joints. The distal joints employ flexible materials to reduce accidental collisions during apple picking. The success rate of the mechanical hand for apple picking was reported as 84%, with an average picking time of 6.0 s per apple. Hohimer et al. [32] created a soft robotic apple harvesting end effector composed of three highly compliant pneumatic actuators arranged symmetrically around a soft, flexible palm, spaced approximately 120° apart in a circle. The separation success rate of apples in experiments was 67%, with an average time from separation to storage of 7.3 s per fruit. Dischinger et al. [17] developed an apple picking end effector incorporating a framework for detecting key events during the picking cycle using hand-held sensors integrated into the fingertips to identify initial grasping quality, fruit slippage, fruit-tree separation, and fruit release from the hand. Zhang et al. [77, 79, 84] designed and tested a flexible swallowing gripper for apple picking, which effectively sensed grasping force, with the mid-finger showing the best force sensing precision. The average absolute error and relative error were 0.153 N and 5.65%, respectively. When grasping with the finger bottoms, better enveloping capability and greater grasping force were obtained, with maximum bending angle and maximum contact force of 22.6° and 5.72 N, respectively. Wang et al. [71] devised a negative pressure four-finger flexible end effector, ensuring contact area with the target fruit on a simple structure basis. They proposed and implemented simultaneous twisting and pulling motions to separate apples from the tree. The compact robot hand structure suited apple grasping and generated considerable grasping force. The method was validated in a natural orchard, with separation rate, damage rate, and harvest rate being 75.6%, 4.55%, and 70.77%, respectively. Zhang et al. [77, 79, 84] and others developed end effectors with soft characteristics and improved cup geometry, enabling flexible adaptation to different contours and orientations of apples. This adaptability ensured efficient grasping while reducing the risk of fruit bruising. Moreover, unlike existing vacuum-based harvesting systems, the use of a compact end effector combined with small vacuum tubes (inner diameter 44.5 mm) to grasp fruits was different. This design choice minimized the chance of sucking leaves or other components into the vacuum chamber. Additionally, a major advantage of using vacuum-based end effectors is their tolerance for some proximity inaccuracies. Operating under a current of vacuum flow, it can attract fruits within approximately 1.5 cm distance, allowing the robotic hand to grab the fruit even if it is not precisely positioned (Fig. 5).

4 Apple Picking Robot Applications and Challenges

4.1 Current Status of Apple Picking Robot Research

The apple-picking robot utilizes computer vision technology for the identification and localization of apples. It feeds the acquired apple information back to the apple-picking robot. Based on this information, the robot determines the optimal picking route and actions. Finally, it utilizes the end effector to separate the fruit, thus completing the apple-picking task. This process seamlessly integrates computer vision, robot control, and end effector technologies to enable automated apple harvesting.

Baenten et al. [4] developed an apple harvesting robot comprising a 6-degree-of-freedom robotic arm and a silicone funnel-shaped vacuum end effector. Equipped with a palm camera at the center, it significantly improved recognition accuracy. Orchard trials showed an 80% success rate in apple picking, with each apple picked in a cycle of 8–10 s. Zhao et al. [85] designed a joint-type picking robotic arm consisting of five rotating joints, with a spoon-shaped end effector covered in rubber sponge

Fig. 5 End-effectors: **a** Silwal et al. [57] developed apple harvesting robot, **b** Hohimer et al. [32] developed soft robotic end-effector for apple harvesting, **c** Dischinger et al. [17] developed end-effector for apple harvesting, **d** Zhang et al. [77, 79, 84] bionic flexible apple gripper, **e** Wang et al. [71] designed pneumatic gripping end-effector and **f** Zhang et al. [77, 79, 84] developed negative pressure end-effector

material and pressure sensors to prevent fruit damage due to excessive clamping force. In orchard trials, the average picking time for this picking robot was 15.4 s, with a success rate of 77%. Zhang et al. [78] devised a pneumatic/electric hybrid-driven three-degree-of-freedom robotic arm. A vacuum end effector was used to adhere to apples, and the robot separated them through rotation and pulling. The success rate was 82.5%, with an average picking time of approximately 8.8 s per apple. Bu et al. [90] integrated a visual system and control system modules into an apple harvesting robot based on industrial robotic arms. They proposed a "pull-bend" picking and harvesting action inspired by human hand-picking actions and conducted picking experiments in orchard environments. In field experiments, the success rate of apple picking was 82.93%, with the entire picking cycle process taking 12.53 ± 0.53 s. Hu et al. [33] designed and built an apple harvesting robot prototype integrated with a binocular camera, a 4-degree-of-freedom robotic hand, a vacuum-based end effector, and a mobile vehicle. In field orchards and simulated orchard environments, the success rate of picking using a rotation-pull mode was 47.37%, with a picking cycle time of < 4 s. Wang et al. [71] designed a negative pressure four-finger flexible end effector and implemented simultaneous twisting-pulling motion to separate apples from trees. The end effector was verified in natural orchards, with separation rate, damage rate, and harvest rate of 75.6%, 4.55%, and 70.77%, respectively. In 2023, Feng et al. [19] developed an apple four-arm picking robot that achieved spatial positioning of target centroids in complex backgrounds and ensured efficient and parallel picking by multiple actuators. Orchard field tests showed varying success rates of harvesting from 71.28 to 80.45%, with average cycle times ranging from 5.8

to 6.7 s. Zhang et al. [77, 79, 84] integrated perception components, a 4-degree-of-freedom robotic hand, a vacuum-based soft end effector, and a dropping component. Field tests demonstrated that the robot system could pick 82.4% of target fruits in orchards with fewer leaf and branch obstructions and successfully harvested 65.2% of target fruits in more challenging orchards with dense branches and leaves. The average picking cycle time of the robot system was approximately 6 s per fruit.

Research on agricultural harvesting robots is advancing towards greater intelligence, efficiency, and reliability.

4.2 *Commercially Available Apple Harvesting Robot*

Abundant Robotics, based in the United States, developed a harvesting robot utilizing parallel and vacuum suction systems [8]. During harvesting, the vacuum head approaches the apple, subjecting it to strong lateral suction force, causing the apple to break at the stem. This reduces contact damage, and the system does not require high precision in apple positioning, greatly improving harvesting efficiency. However, this suction method necessitates powerful negative pressure support equipment, resulting in a bulky harvesting robot.

Israeli company FFRobotics designed and developed a fully automated apple harvesting robot [14], featuring 12 telescopic robotic arms arranged in two rows of six arms each. The harvested apples are placed on a series of soft rubber conveyors for packing, as illustrated in the figure. By utilizing a set of cameras to identify apples, all 12 arms can harvest simultaneously, resulting in higher efficiency compared to single-arm systems. In structured orchards, it can harvest up to 8000 apples per hour.

Tevel Aerobotics introduced an autonomous flying fruit-picking robot [15], consisting of a wheeled base drone and four quadcopters tethered to the base. The drones employ integrated gripper arms with end effectors. The suggested apple harvesting end effector is a vacuum suction cup, however, further details or evaluations are not provided. This system is suitable for apples as well as other similarly sized fruits such as oranges and mangoes.

4.3 *Problems Facing Apple's Harvesting Robots*

(1) Harvesting speed is slow. In complex natural environments, the network structure is relatively complex, while the computational capacity of apple-picking robots is limited, leading to low real-time performance of the vision system. Most apple-picking robots use a single end effector, requiring a "reach," "pluck," and "retrieve" process for each harvest, resulting in long harvesting cycles for individual fruits. Current path planning also faces challenges such as difficulty

in environmental modeling, slow algorithm convergence speed, and susceptibility to local optima. Additionally, the separation of algorithm perception and decision-making makes it difficult to apply in practical environments.

(2) Low harvesting success rate. Harvesting apples in complex natural environments is affected by factors such as lighting conditions, fruit overlap, and foliage obstruction, leading to misidentification or missed detection. Wind or apple oscillations caused by harvesting can introduce significant errors between the positioning results and the actual situation, affecting positioning accuracy and causing failures in end effector grasping. Even if apples obstructed by branches and leaves are successfully identified and located, the mechanical arm and end effector may not be able to approach the target fruit due to harvesting path restrictions.

(3) Apple damage occurs. Apple skin is fragile and prone to damage during mechanized harvesting operations, resulting in decreased fruit quality. The main causes of fruit damage include errors in identification and positioning, which cause the end effector to collide with the fruit; excessive force applied during harvesting causing fruit bruising; and current apple identification and positioning algorithms often only determine the center of the apple without recognizing its orientation, preventing the end effector from rotating or stretching the fruit relative to the stem in a specific manner, leading to fruit damage.

(4) High costs. Due to the complex working environment and high requirements for vision systems, mechanical arms, end effectors, and control systems, apple-picking robots, mostly employing collaborative robotic arms, incur high production costs. Apple harvesting is seasonal work, resulting in low utilization rates for apple-picking robots, high operating costs, and low profitability.

5 Future Trends

(1) Reduce the complexity of orchard environments. Clear visibility and separation of fruits can reduce the difficulty of recognition for deep learning networks and the task of fruit picking for end effectors. Future efforts will focus on tree pruning and thinning direction to further improve the working environment, enhance harvesting efficiency, and advance the development of mechanized apple picking.

(2) Algorithm optimization. To further enhance the efficiency of apple-picking robots in natural environments, deep learning algorithms need further optimization. This can be achieved by enhancing datasets, improving feature extraction modules, adding attention mechanisms, and implementing pruning, distillation, and other methods to improve the accuracy or speed of apple recognition. Additionally, future research will focus more on the three-dimensional pose estimation of apples to ensure that end effectors can rotate or stretch relative to the stem in specific ways, thereby increasing separation success rates and reducing apple

damage. Intelligent path planning algorithms also need further improvement to enhance picking efficiency when dealing with dynamic environments.

(3) Multi-sensor fusion. To address the current issue of depth distortion with single sensors, multiple sensors can be integrated for apple picking. Pressure sensors, palm cameras, and other sensors can be integrated into end effectors to improve positioning accuracy and ensure non-destructive picking of apples, enhancing the adaptability of apple-picking robots to complex operational environments.

(4) Multi-arm cooperation. While research on single-arm picking technology has advanced, it is limited by picking speed compared to manual picking. Moreover, the working space of a single mechanical arm is fixed and may not align well with the distribution of fruit on apple trees. Utilizing multiple mechanical arms can better align with the distribution of fruit, avoiding wastage of picking space, significantly improving the picking efficiency of individual apple-picking robots, and even exceeding human picking speeds.

6 Conclusions

The task of apple harvesting, being labor-intensive and time-constrained, urgently calls for mechanization. This paper provides a review of the current status of research and application of apple harvesting robots, identifying areas such as harvesting efficiency, minimizing damage, and reducing costs where significant improvements are still needed. Addressing the existing challenges, suggestions are proposed, including reducing orchard complexity, optimizing recognition and planning algorithms, integrating multiple sensors, utilizing pneumatic-driven end effectors, and fostering multi-arm collaboration. These recommendations aim to accelerate the innovation and development of key technologies in future apple harvesting robot research. The ultimate goal is to design a commercially viable apple harvesting robot that is adaptable to complex environments, efficient, low-damage, cost-effective, easy to maintain, fully functional, and user-friendly. This endeavor holds significant practical importance in achieving automation in apple harvesting.

Acknowledgements This research is supported by the National Key Research and Development Program of China (2023YFE0122600). The experiment was conducted at Beijing Shengshi Fruit King Agricultural Technology Co., LTD.

References

1. Andriyanov N, Khasanshin I, Utkin D, Gataullin T, Ignar S, Shumaev V, Soloviev V (2022) Intelligent system for estimation of the spatial position of apples based on yolov3 and real sense depth camera d415. Symmetry 14(1):148. https://doi.org/10.3390/sym14010148
2. Anwar I, Lee S (2017) High performance stand-alone structured light 3d camera for smart manipulators. IEEE 192–195. https://doi.org/10.1109/URAI.2017.7992709

3. Atif M, Lee S (2018) Adaptive pattern resolution for structured light 3d camera system. IEEE 1–4. https://doi.org/10.1109/ICSENS.2018.8589640
4. Baeten J, Donné K, Boedrij S, Beckers W, Claesen E (2008) Autonomous fruit picking machine: a robotic apple harvester. In: Laugier C, Siegwart R (eds) Field and service robotics: results of the 6th international conference. Springer Berlin Heidelberg, Berlin, Heidelberg, pp 531–539. https://doi.org/10.1007/978-3-540-75404-6_51
5. Boryga M, Graboś A, Kołodziej P, Gołacki K, Stropek Z (2015) Trajectory planning with obstacles on the example of tomato harvest. Agric Agric Sci Proc 7:27–34. https://doi.org/10.1016/j.aaspro.2015.12.026
6. Bu LX, Chen CK, Hu GR, Sugirbay A, Chen J (2020) Technological development of robotic apple harvesters: a review. Inmateh Agric Eng 61(2). https://doi.org/10.35633/inmateh-61-17
7. Bulanon DM, Kataoka T, Ota Y, Hiroma T (2002) Ae—automation and emerging technologies: a segmentation algorithm for the automatic recognition of fuji apples at harvest. Biosyst Eng 83(4):405–412. https://doi.org/10.1006/bioe.2002.0132
8. Burwood-Taylor L (2017) Abundant robotics raises $10m series A for apple picking robot led by GV. https://agfundernews.com/abundant-robotics-raises-10m-series-a-for-apple-picking-robot
9. Cao X, Zou X, Jia C, Chen M, Zeng Z (2019) RRT-based path planning for an intelligent litchi-picking manipulator. Comput Electron Agric 156:105–118. https://doi.org/10.1016/j.compag.2018.10.031
10. Ceccarelli M, Criado EJ, Figliolini G, Mata AS, Ottaviano E (2000) Designing a robotic gripper for harvesting horticulture products. Robotica 18(1):105–111. https://doi.org/10.1017/S026357479900226X
11. Chen B, Gong L, Yu C, Du X, Chen J, Xie S, Le X, Li Y, Liu C (2023) Workspace decomposition based path planning for fruit-picking robot in complex greenhouse environment. Comput Electron Agric 215:108353. https://doi.org/10.1016/j.compag.2023.108353
12. Chen W, Zhang J, Guo B, Wei Q, Zhu Z, Tang Y (2021) An apple detection method based on des-yolo v4 algorithm for harvesting robots in complex environment. Math Probl Eng 2021:7351470. https://doi.org/10.1155/2021/7351470
13. Chu P, Li Z, Lammers K, Lu R, Liu X (2021) Deep learning-based apple detection using a suppression mask R-CNN. Pattern Recogn Lett 147:206–211. https://doi.org/10.1016/j.patrec.2021.04.022
14. Courtney R, Mullinax TJ (2019) Washington orchards host robotic arms race—video. https://www.goodfruit.com/washington-orchards-host-robotic-arms-race/
15. Crumley B (2022) Apple drones are now reality… in orchards cultivating fruit. https://dronedj.com/2022/01/31/apple-drones-are-now-reality-in-orchards-cultivating-fruit/
16. D'Esnon AG (1985) Robotic harvesting of apples. Agri-Mation 1:210–214
17. Dischinger LM, Cravetz M, Dawes J, Votzke C, VanAtter C, Johnston ML, Grimm CM, Davidson JR (2021) Towards intelligent fruit picking with in-hand sensing. IEEE 3285–3291. https://doi.org/10.1109/IROS51168.2021.9636341
18. Feng J, Zeng L, He L (2019) Apple fruit recognition algorithm based on multi-spectral dynamic image analysis. Sensors 19(4):949. https://doi.org/10.3390/s19040949
19. Feng Q, Zhao C, Li T, Chen L, Guo X, Xie F, Xiong Z, Chen K, Liu C, Yan T (2023) Design and test of a four-arm apple harvesting robot. Trans Chin Soc Agric Eng 39(13):25–33. https://doi.org/10.11975/j.issn.1002-6819.202305114
20. Gai R, Guo Y (2021) Research on acceleration and deceleration control algorithm of s-curve. IEEE 2104–2110. https://doi.org/10.1109/HPCC-DSS-SmartCity-DependSys53884.2021.00313
21. Gai R, Wang X, Chang Z, Guo Y (2022) Fruit and vegetable picking robot movement planning: a review. In: 2022 IEEE international conferences on internet of things (iThings) and IEEE green computing & communications (GreenCom) and IEEE cyber, physical & social computing (CPSCom) and IEEE smart data (SmartData) and IEEE congress on cybermatics (Cybermatics), pp 379–384. https://doi.org/10.1109/iThings-GreenCom-CPSCom-SmartData-Cybermatics55523.2022.00086

22. Gallardo RK, Brady MP (2015) Adoption of labor-enhancing technologies by specialty crop producers: the case of the Washington apple industry. Agric Financ Rev 75(4):514–532
23. Gao F, Fu L, Zhang X, Majeed Y, Li R, Karkee M, Zhang Q (2020) Multi-class fruit-on-plant detection for apple in snap system using faster R-CNN. Comput Electron Agric 176:105634. https://doi.org/10.1016/j.compag.2020.105634
24. Gao M, Ding P, Yang Y (2015) Time-optimal trajectory planning of industrial robots based on particle swarm optimization. IEEE 1934–1939. https://doi.org/10.1109/IMCCC.2015.410
25. Gao R, Zhou Q, Cao S, Jiang Q (2022) An algorithm for calculating apple picking direction based on 3d vision. Agriculture 12(8):1170. https://doi.org/10.3390/agriculture12081170
26. Gené-Mola J, Gregorio E, Guevara J, Auat F, Sanz-Cortiella R, Escolà A, Llorens J, Morros J, Ruiz-Hidalgo J, Vilaplana V, Rosell-Polo JR (2019) Fruit detection in an apple orchard using a mobile terrestrial laser scanner. Biosyst Eng 187:171–184. https://doi.org/10.1016/j.biosystemseng.2019.08.017
27. Gené-Mola J, Sanz-Cortiella R, Rosell-Polo JR, Morros J, Ruiz-Hidalgo J, Vilaplana V, Gregorio E (2020) Fruit detection and 3d location using instance segmentation neural networks and structure-from-motion photogrammetry. Comput Electron Agric 169:105165. https://doi.org/10.1016/j.compag.2019.105165
28. Gongal A, Amatya S, Karkee M, Zhang Q, Lewis K (2015) Sensors and systems for fruit detection and localization: a review. Comput Electron Agric 116:8–19. https://doi.org/10.1016/j.compag.2015.05.021
29. Guan Y, Yokoi K, Stasse O, Kheddar A (2005) On robotic trajectory planning using polynomial interpolations. IEEE 111–116. https://doi.org/10.1109/ROBIO.2005.246411
30. He L, Fu H, Sun D, Karkee M, Zhang Q (2017) Shake-and-catch harvesting for fresh market apples in trellis-trained trees. Trans ASABE 60(2):353–360. https://doi.org/10.13031/trans.12067
31. He L, Zhang X, Ye Y, Karkee M, Zhang Q (2019) Effect of shaking location and duration on mechanical harvesting of fresh market apples. Appl Eng Agric 35(2):175–183. https://doi.org/10.13031/aea.12974
32. Hohimer CJ, Wang H, Bhusal S, Miller J, Mo C, Karkee M (2019) Design and field evaluation of a robotic apple harvesting system with a 3d-printed soft-robotic end-effector. Trans Asabe 62(2):405–414. https://doi.org/10.13031/trans.12986
33. Hu G, Chen C, Chen J, Sun L, Sugirbay A, Chen Y, Jin H, Zhang S, Bu L (2022) Simplified 4-dof manipulator for rapid robotic apple harvesting. Comput Electron Agric 199:107177. https://doi.org/10.1016/j.compag.2022.107177
34. Hua Y (2021) Review of trajectory planning for industrial robots. J Phys Conf Ser 1906:12041. https://doi.org/10.1088/1742-6596/1906/1/012041
35. Ji W, Zhao D, Cheng F, Xu B, Zhang Y, Wang J (2012) Automatic recognition vision system guided for apple harvesting robot. Comput Electr Eng 38(5):1186–1195. https://doi.org/10.1016/j.compeleceng.2011.11.005
36. Jia W, Mou S, Wang J, Liu X, Zheng Y, Lian J, Zhao D (2020) Fruit recognition based on pulse coupled neural network and genetic Elman algorithm application in apple harvesting robot. Int J Adv Robot Syst 17(1):1737054849. https://doi.org/10.1177/1729881419897473
37. Jia W, Tian Y, Luo R, Zhang Z, Lian J, Zheng Y (2020) Detection and segmentation of overlapped fruits based on optimized mask R-CNN application in apple harvesting robot. Comput Electron Agric 172:105380. https://doi.org/10.1016/j.compag.2020.105380
38. Jiang G, Zhao C (2012) Apple recognition based on machine vision. IEEE 1148–1151. https://doi.org/10.1109/ICMLC.2012.6359517
39. Kang H, Wang X, Chen C (2022) Accurate fruit localisation using high resolution Lidar-camera fusion and instance segmentation. Comput Electron Agric 203:107450. https://doi.org/10.1016/j.compag.2022.107450
40. Li T, Qiu Q, Zhao C, Xie F (2021) Task planning of multi-arm harvesting robots for high-density dwarf orchards. Trans Chin Soc Agric Eng 37(2):1–10. https://doi.org/10.11975/j.issn.1002-6819.2021.2.001

41. Li T, Fang W, Zhao G, Gao F, Wu Z, Li R, Fu L, Dhupia J (2023) An improved binocular localization method for apple based on fruit detection using deep learning. Inf Process Agric 10(2):276–287. https://doi.org/10.1016/j.inpa.2021.12.003
42. Li T, Xie F, Zhao Z, Zhao H, Guo X, Feng Q (2023) A multi-arm robot system for efficient apple harvesting: perception, task plan and control. Comput Electron Agric 211:107979. https://doi.org/10.1016/j.compag.2023.107979
43. Li Y, Feng Q, Zhang Y, Peng C, Ma Y, Liu C, Ru M, Sun J, Zhao C (2024) Peduncle collision-free grasping based on deep reinforcement learning for tomato harvesting robot. Comput Electron Agric 216:108488. https://doi.org/10.1016/j.compag.2023.108488
44. Lu S, Chen W, Zhang X, Karkee M (2022) Canopy-attention-yolov4-based immature/mature apple fruit detection on dense-foliage tree architectures for early crop load estimation. Comput Electron Agric 193:106696. https://doi.org/10.1016/j.compag.2022.106696
45. Luo T, Escalante CL (2017) Us farm workers: what drives their job retention and work time allocation decisions? Econ Labour Relat Rev 28(2):270–293. https://doi.org/10.1177/1035304617703933
46. Lv J, Ni H, Wang Q, Yang B, Xu L (2019) A segmentation method of red apple image. Sci Hortic 256:108615. https://doi.org/10.1016/j.scienta.2019.108615
47. Ma X, Liu G, Zhou W, Feng J (2013) Apple recognition based on fuzzy neural network and quantum genetic algorithm. Trans Chin Soc Agric Machi 44(12):227–232. https://doi.org/10.6041/j.issn.1000-1298.2013.12.038
48. Mahalakshmi S, Arokiasamy A (2021) A new particle swarm optimisation variant-based experimental verification of an industrial robot trajectory planning. Int J Oper Res 41(3):399–422. https://doi.org/10.1504/IJOR.2021.116250
49. Mhamed M, Zhang Z, Yu J, Li Y, Zhang M (2024) Advances in apple's automated orchard equipment: a comprehensive research. Comput Electron Agric 221:108926. https://doi.org/10.1016/j.compag.2024.108926
50. Nguyen T, Vandevoorde K, Wouters N, Kayacan E, De Baerdemaeker J, Saeys W (2016) Detection of red and bicoloured apples on tree with an RGB-D camera. Biosyst Eng 146. https://doi.org/10.1016/j.biosystemseng.2016.01.007
51. Peterson DL, Miller SS (1989) Advances in mechanical harvesting of fresh market quality apples. J Agric Eng Res 42(1):43–50. https://doi.org/10.1016/0021-8634(89)90038-3
52. Peterson DL, Wolford SD (2003) Fresh–market quality tree fruit harvester part ii: apples. Appl Eng Agric 19(5):545. https://doi.org/10.13031/2013.15314
53. Porawagama CD, Munasinghe SR (2014) Reduced jerk joint space trajectory planning method using 5-3-5 spline for robot manipulators. IEEE 1–6. https://doi.org/10.1109/ICIAFS.2014.7069580
54. Sa I, Ge Z, Dayoub F, Upcroft B, Perez T, McCool C (2016) Deepfruits: a fruit detection system using deep neural networks. Sensors 16(8):1222. https://doi.org/10.3390/s16081222
55. Sarig Y (1993) Robotics of fruit harvesting: a state-of-the-art review. J Agric Eng Res 54(4):265–280. https://doi.org/10.1006/jaer.1993.1020
56. Si Y, Liu G, Feng J (2015) Location of apples in trees using stereoscopic vision. Comput Electron Agric 112:68–74. https://doi.org/10.1016/j.compag.2015.01.010
57. Silwal A, Davidson JR, Karkee M, Mo C, Zhang Q, Lewis K (2017) Design, integration, and field evaluation of a robotic apple harvester. J. Field Robot 34(6):1140–1159. https://doi.org/10.1002/rob.21715
58. Song J (2012) Target positioning of picking robot fusing laser ranging and vision. In: Yang Y, Ma M (eds) Green communications and networks: proceedings of the international conference on green communications and networks (GCN 2011), vol 113. Springer Netherlands, Dordrecht 113, pp 985–993. https://doi.org/10.1007/978-94-007-2169-2_117
59. Suchet B, James U (2017) Deep fruit detection in Orchards. In: 2017 IEEE international conference on robotics and automation (ICRA), pp 3626–3633. https://doi.org/10.1109/ICRA.2017.7989417
60. Subeesh A, Mehta CR (2021) Automation and digitization of agriculture using artificial intelligence and internet of things. Artif Intell Agric 5:278–291. https://doi.org/10.1016/j.aiia.2021.11.004

61. Sun M, Xu L, Luo R, Lu Y, Jia W (2022) Fast location and recognition of green apple based on RGB-D image. Front Plant Sci 13. https://doi.org/10.3389/fpls.2022.864458
62. Tabb A, Peterson D, Park J (2006) Segmentation of apple fruit from video via background modeling. ASABE 063060. https://doi.org/10.13031/2013.20873
63. Takahashi T, Zhang S, Fukuchi H (2002) Acquisition of 3-d information by binocular stereo vision for vehicle navigation through an Orchard. ASABE, pp 337–346. https://doi.org/10.13031/2013.10023
64. Tang X, Zhou H, Xu T (2024) Obstacle avoidance path planning of 6-dof robotic arm based on improved a* algorithm and artificial potential field method. Robotica 42(2):457–481. https://doi.org/10.1017/S0263574723001546
65. Tian Y, Yang G, Wang Z, Wang H, Li E, Liang Z (2019) Apple detection during different growth stages in orchards using the improved yolo-v3 model. Comput Electron Agric 157:417–426. https://doi.org/10.1016/j.compag.2019.01.012
66. Tsoulias N, Paraforos DS, Xanthopoulos G, Zude-Sasse M (2020) Apple shape detection based on geometric and radiometric features using a lidar laser scanner. Remote Sens 12(15):2481. https://doi.org/10.3390/rs12152481
67. Tu S, Pang J, Liu H, Zhuang N, Chen Y, Zheng C, Wan H, Xue Y (2020) Passion fruit detection and counting based on multiple scale faster R-CNN using RGB-D images. Precision Agric 21:1072–1091. https://doi.org/10.1007/s11119-020-09709-3
68. USDA (2023) Fresh apples, grapes, and pears: world markets and trade. Foreign Agricultural Service. https://apps.fas.usda.gov/psdonline/circulars/
69. Wang D, Xu Y, Song H, He D, Zhang H (2015) Fusion of K-means and Ncut algorithm to realize segmentation and reconstruction of two overlapped apples without blocking by branches and leaves. Trans Chin Soc Agric Eng 31(10):227–234. https://doi.org/10.11975/j.issn.1002-6819.2015.10.030
70. Wang H, Wang H, Huang J, Zhao B, Quan L (2019) Smooth point-to-point trajectory planning for industrial robots with kinematical constraints based on high-order polynomial curve. Mech Mach Theory 139:284–293. https://doi.org/10.1016/j.mechmachtheory.2019.05.002
71. Wang X, Kang H, Zhou H, Au W, Wang MY, Chen C (2023) Development and evaluation of a robust soft robotic gripper for apple harvesting. Comput Electron Agric 204:107552
72. Wachs JP, Stern HI, Burks T, Alchanatis V (2010) Low and high-level visual feature-based apple detection from multi-modal images. Precis Agric 11(6):717–735. https://doi.org/10.1007/s11119-010-9198-x
73. Wheat D (2019) Robotic apple pickers make progress. https://www.capitalpress.com/ag_sectors/orchards_nuts_vines/robotic-apple-pickers-make-progress/_b1bd180e-1163-11ea-93ec-bf5cfc2aecf1.html
74. Xie J, Zhang Z, Wei Z, Ma S (2019) Simulation of apple picking path planning based on artificial potential field method. IOP Conf Ser Earth Environ Sci 252(5):52148. https://doi.org/10.1088/1755-1315/252/5/052148
75. Xu J, Gu B, Tian G (2022) Review of agricultural lot technology. Artif Intell Agric 6:10–22. https://doi.org/10.1016/j.aiia.2022.01.001
76. Yuan J (2020) Research progress analysis of robotics selective harvesting technologies. Trans Chin Soc Agric Machi 51:1–17. https://doi.org/10.6041/j.issn.1000-1298.2020.09.001
77. Zhang C, Wang H, Fu L, Pei Y, Lan C, Hou H, Song H (2023) Three-dimensional continuous picking path planning based on ant colony optimization algorithm. PLoS ONE 18(2):e282334. https://doi.org/10.1371/journal.pone.0282334
78. Zhang K, Lammers K, Chu P, Li Z, Lu R (2021) System design and control of an apple harvesting robot. Mechatronics 79:102644. https://doi.org/10.1016/j.mechatronics.2021.102644
79. Zhang K, Lammers K, Chu P, Li Z, Lu R (2023) An automated apple harvesting robot—from system design to field evaluation. J Field Robot. https://doi.org/10.1002/rob.22268
80. Zhang K, Chu P, Lammers K, Li Z, Lu R (2024) Active laser-camera scanning for high-precision fruit localization in robotic harvesting: system design and calibration. Horticulturae 10(1):40. https://doi.org/10.3390/horticulturae10010040

81. Zhang Z, Heinemann P, Liu J, Schupp, J, Baugher T (2016) Design and field test of a low-cost apple harvest-assist unit. Trans. ASABE 59(5):1149–1156. https://doi.org/10.13031/trans.59.11708
82. Zhang Z, Heinemann PH (2017) Economic analysis of a low-cost apple harvest-assist unit. Horttechnology 27(2):240–247. https://doi.org/10.21273/HORTTECH03548-16
83. Zhang Z, Igathinathane C, Li J, Cen H, Lu Y, Flores P (2020) Technology progress in mechanical harvest of fresh market apples. Comput Electron Agric 175:105606. https://doi.org/10.1016/j.compag.2020.105606
84. Zhang Z, Zhou J, Yi B, Zhang B, Wang K (2023) A flexible swallowing gripper for harvesting apples and its grasping force sensing model. Comput Electron Agric 204:107489. https://doi.org/10.1016/j.compag.2022.107489
85. Zhao DA, Lv JD, Ji W, Zhang Y, Chen Y (2011) Design and control of an apple harvesting robot. Biosyst Eng 110(2):112–122. https://doi.org/10.1016/j.biosystemseng.2011.07.005
86. Zheng W, Xie Y, Zhang B, Zhou J, Zhang J (2021) Dexterous robotic grasping of delicate fruits aided with a multi-sensory e-glove and manual grasping analysis for damage-free manipulation. Comput Electron Agric 190:106472. https://doi.org/10.1016/j.compag.2021.106472
87. Zhou H, Wang X, Au W, Kang H, Chen C (2022) Intelligent robots for fruit harvesting: recent developments and future challenges. Precis Agric 23(5):1856–1907. https://doi.org/10.1007/s11119-022-09913-3
88. Zhuang M, Li G, Ding K (2023) Obstacle avoidance path planning for apple picking robotic arm incorporating artificial potential field and a algorithm. IEEE Access 11:100070–100082. https://doi.org/10.1109/ACCESS.2023.3312763
89. Kang H, Chen C (2020) Fast implementation of real-time fruit detection in apple orchards using deep learning. Comput Electron Agric 168:105108. https://doi.org/10.1016/j.compag.2019.105108
90. Bu L, Chen C, Hu G, Sugirbay A, Sun H, Chen J (2022) Design and evaluation of a robotic apple harvester using optimized picking patterns. Comput Electron Agric 198:107092. https://doi.org/10.1016/j.compag.2022.107092

Design, Assembly and Test of a Low-Cost Vacuum Based Apple Harvesting Robot

Wanjia Hua, Wenqiang Zhang, Zhao Zhang, Xiaohang Liu, Chayan Saha, Can Hu, and Xufeng Wang

Abstract The timely harvesting of fresh apples faces challenges due to labor shortages. Robotic harvesting has the potential to alleviate the labor intensity on operators and liberate additional agricultural labor. However, the current success rate of apple-picking robots is low, resulting in fruit damage, high operational costs, and an inability to meet market demand. In this research, a prototype apple harvesting manipulator was designed and constructed, providing a solid foundation for future advancements. The manipulator prototype includes a vacuum three-rotational (3R) degrees-of-freedom (DoF) end-effector, a 3-DoF Cartesian system, an RGB-D camera, and a mobile vehicle. Vision-assisted positioning system and manipulator controller were designed to achieve precise positioning and separation motions of the manipulator. The harvesting experiment was conducted in a dwarf, densely planted spindle apple orchard. Results from the experiment indicate that the rotation followed by pull mode proved more effective than the pull mode for apple picking. In the critical negative pressure environment of − 65 kPa employed for apple picking, utilizing 33 mm suction cups has demonstrated an 81.7% separation success rate. The diameter center as the optimal theoretical point for adsorption with an allowable two-dimensional error mean value of 9.62%. The manipulator ensures essentially damage-free picking of target fruits, demonstrating a stem damage rate of 17.1%. The developed picking

---

W. Hua · W. Zhang
China Agricultural University, Beijing 100083, China

Z. Zhang (✉) · X. Liu
Key Laboratory of Smart Agriculture System Integration, Ministry of Education, Beijing 100083, China
e-mail: zhaozhangcau@cau.edu.cn

Key Laboratory of Agricultural Information Acquisition Technology, Ministry of Agriculture and Rural Affairs, China Agricultural University, Beijing 100083, China

C. Saha
Department of Farm Power and Machinery, Bangladesh Agricultural University, Mymensingh 2202, Bangladesh

C. Hu · X. Wang
College of Mechanical and Electronic Engineering, Tarim University, Alar 843300, Xinjiang, China

Y. Liu et al. (eds.), *New Technologies Applied in Apple Production*, Smart Agriculture 10, https://doi.org/10.1007/978-981-97-7778-5_2

prototype successfully accomplishes the task of apple harvesting, boasting main advantages of compact size, low cost, non-destructiveness, and high flexibility.

Keywords Fruit harvesting · Vacuum suction · Agricultural robot · Manipulator · Field evaluation

1 Highlights

- The design and development of a manipulator for harvesting apples is presented.
- Manipulator controller and vision positioning systems were designed.
- – 65 kPa is the critical negative pressure for picking with a 33 mm vacuum sucker.
- The equatorial center of the apple is the best attachment point.

2 Introduction

Fresh apple harvesting is labor-intensive [11, 18], primarily relying on manual labor. This manual approach is associated with inefficiency, high costs, and a heightened risk of occupational diseases [9, 38]. Furthermore, the strain on labor resources is increasing in the face of labor shortages, escalating labor costs, and an aging workforce [21, 37]. The challenges of harvesting ripe fruit in a timely, efficient, and cost-effective manner are increasingly evident, posing a threat to the sustainable development of the apple industry. Hence, developing mechanized harvesting techniques that can replace manual labor has become imperative [26].

Currently, mechanized apple harvesting technology is categorized into semi-automatic harvesting technology, harvest-assisted platforms, and robot harvesting technology [33]. In semi-automatic harvesting techniques, batch harvesting is conducted through a shaking picking method [6, 14, 15], which encounters challenges in preventing damage [23, 24]. The utilization of harvest-assisted platforms stands out as a primary approach for current commercialization [35, 36]. Nevertheless, its failure to address the core issue of labor shortage, and lack of compatibility with various orchard structures pose significant challenges [2]. Robotic harvesting has an opportunity to reduce labor intensity of operators and minimize fruit damage, hence streamlining agricultural production costs and meeting timely harvesting operations expectations [27, 28, 30], making it the predominant trajectory in current mechanized harvesting trends [38]. Since the 1970s, researchers have been developing robots for fruit harvesting [4, 12, 25]. France pioneered in 1984 by initiating research on apple harvesting robots [13, 22]. However, progress in developing apple harvesting robots has been gradual, primarily due to the challenges posed by unstructured environments and complex disturbance conditions in the field [31].

With the groundbreaking advancements in standardized orchard production technology and artificial intelligence, recent research has seen pioneering efforts in developing autonomous intelligent apple-harvesting robots. Based in the USA, Abundant Robotics has engineered a harvesting robot employing a parallel connection and vacuum intake mechanism. This robotic picking solution demonstrates high efficiency in harvesting [5]. However, it faces challenges due to its bulky structure, preventing effective penetration into the canopy. Additionally, this results in issues like high energy consumption, elevated noise levels, and a propensity for causing damage to the fruits. Silwal et al. introduced an apple harvesting robot featuring a cable-driven three-finger robotic gripper designed to detach apples from trees efficiently. This robotic system demonstrated an overall success rate of 84% with a cycle time of 6 s per apple [26]. The inadvertent entanglement of both branches and fruit during the gripping action can fail in the apple grip or even damage to the robotic gripper. Zhang et al. employed a three-degree-of-freedom manipulator and a vacuum-based end-effector for apple separation, achieving a picking success rate of 64.06% with an average cycle time of approximately 8.8 s [34]. This system is notable for its swift picking speed and the capability to pick apples in various orientations. Nevertheless, it has certain limitations, such as a confined working range. Moreover, the size of the vacuum suction cup may lead to unintended leaf intake, resulting in the suction tube being prone to jamming during operation. Hu et al. developed an apple harvesting robot featuring a 4-DoF manipulator and a vacuum-based end-effector. The harvesting cycle time was less than 4 s, and the rate of apple stem damage was 11.11%, meeting commercialization requirements [17]. However, owing to the limited flexibility of the 4-DoF manipulator, the end-effector struggled to approach and separate target fruits in diverse orientations, making it challenging to execute complex picking patterns. Consequently, the picking success rate was only 47.37%. Wang et al. introduce a new soft robotic gripper consisting of four tapered soft robotic fingers (SRF) and a multi-mode suction cup [28]. Highly adaptive bending motions can be generated from predefined geometries [7]. The gripper achieves a detachment rate of 75.6%, damage rate of 4.55%, and harvesting rate of 70.77%. On average, it takes 14.69 s to harvest one apple.

Aiming at the problems of high cost, long picking cycle, insufficient adaptability and inflexible movement of existing manipulators, the main contributions of the thesis are as follows: Firstly, a prototype of a 6-DOF picking manipulator has been designed and constructed, featuring a simple and compact structural design aimed at reducing costs and minimizing fruit damage rates. Secondly, leveraging a kinematic model, a manipulator controller was developed that collaborates with a vision system for efficient execution of fruit-picking tasks. Finally, field experiments were conducted to assess the prototype's performance, laying a solid foundation for future research endeavors.

In Sect. 2, the paper outlines the system design and manipulator control methodology of the apple-picking robot. Section 3 provides an in-depth presentation of the results obtained from field picking experiments. The paper concludes with Sect. 4, summarizing key findings and drawing conclusions.

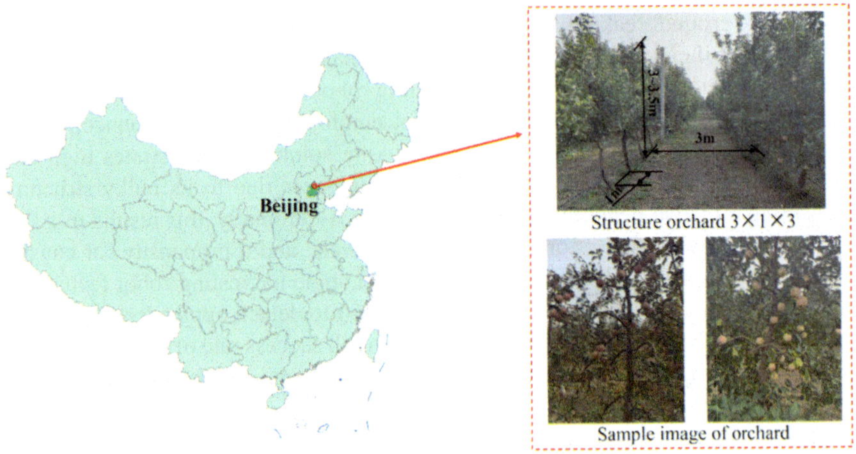

Fig. 1 Growing environment for apple fields

3 Materials and Methods

3.1 The Orchard Standards and Apple Tree Structures

The Design of standardized apple orchards in China adopts a dwarf, densely planted spindle tree form. This Design offers several key advantages, including compactness, open space within rows, improved ventilation and light penetration, and enhanced suitability for mechanization. Specifically, the fruit trees in this Design feature a row spacing of approximately 3 m, with a plant spacing of 1 m. The main trunk's height ranges from 3.0 to 3.5 m (refer to Fig. 1). Around the main trunk, 10–15 small main branches are growing in a staggered manner, while the main branches are left devoid of lateral branches and extend directly to bear fruit. Overall, the tree has a wider base and a smaller top, giving it a broadly conical form. Because of the inter-row surface's relative flatness, orchard equipment has plenty of room to work. Fruits are mostly found in the height range of 0.5–2.0 m, mostly growing on the main branches.

3.2 System Design

The main principles of hardware design for apple picking manipulators are non-destructive, fast and flexible movement, adaptability and low cost. Efficient actuators and drive systems are used to achieve fast and accurate positioning. Flexibility of movement is achieved through articulated design. Soft materials and a proper separation pattern ensure that the robot avoids damaging the fruit during the picking

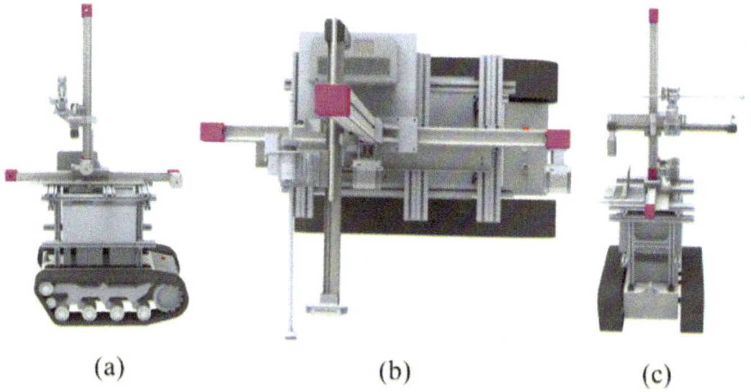

Fig. 2 **a** Front view of apple picking prototype, **b** left view, and **c** top view

process. Figure 2 depicts the apple harvesting system, comprising four key components: a 3R DoF vacuum end-effector, a 3-DoF Cartesian system, an RGB-D camera, and a mobile vehicle.

The vacuum end-effector plays a pivotal role in the manipulator and is responsible for gripping and separating the fruit. Its 3R Degree of Freedom (DoF) configuration enables adaptability to various angles and orientations, ensuring subsequent efficient and precise picking. The 3-DoF Cartesian system enables the manipulator to move quickly through three-dimensional space, allowing for precise positioning when approaching fruit. This arm design with fewer degrees of freedom improves the manipulator's adaptability to the orchard environment, making it more suitable for efficient harvesting. The Realsense D435i was employed to aid in the precise positioning of apples. Throughout the operation, the 3-DoF Cartesian system was positioned along one side of the rows of fruit trees. Meanwhile, the vacuum end-effector was securely affixed to the end of the manipulator, and the camera, oriented in the same direction as the manipulator, was horizontally mounted at the end of the y-axis. To achieve versatile maneuverability between rows of fruit trees, the entire manipulator system was installed on a mobile vehicle equipped with a tracked chassis (MID-01, Shenzhen Yuhosun Technology Co., Ltd.).

All the electronic components of the entire system are integrated into the mobile vehicle, encompassing the control cabinet, vacuum pump, vacuum solenoid valve, mobile lithium battery, power converter, and other essential elements. The overall compactness and portability of the system is ensured. The control process is executed through an integrated graphical user interface (GUI), allowing for intuitive and flexible monitoring and adjustment of the entire harvesting process.

3.2.1 End-Effector Design

The end-effector serves as a pivotal component of the robot, tasked with executing its assigned functions. As the sole system element in direct physical contact with the fruit, the end-effector significantly influences harvesting performance. In the context of an apple harvesting robot, the end-effector faces challenges arising from the dense fruit tree environment. The unpredictable growth of branches poses a challenge for the end-effector to reach the target location in the desired direction. To achieve precise positioning of the end-effector in a specific orientation and minimize space utilization, a 3R DoF design, incorporating pitch, roll, and yaw movements, was implemented.

The vacuum end-effector (Fig. 3) was designed for non-destructive apple harvesting and consisted mainly of three rotating joints and a hollow aluminum tube. A vacuum pump (Type: PCV-4 M, made by Fujiwara Co., Ltd., China) is chosen as the vacuum generator to enable grasping, and a corrugated vacuum sucker is used as the gripper. The vacuum pump has a maximum vacuum of 20 Pa and a rated pumping speed of 100 L/min. It rated pumping speed is 100 L/min, and its maximum vacuum is 20 Pa. With the vacuum pump attached to the vacuum suction cup via an aluminum tube, the vacuum solenoid valve may be switched on and off more easily. This allows the fruit to be attached or released. When the vacuum pump operates, it evacuates air from the pipe to create suction, causing the suction cups to adhere to the surface of the apple firmly. The hollow tube extends the reach at the end while also serving as an air storage tank.

After the manipulator successfully grips the target fruit, the choice of an appropriate separation strategy becomes a pivotal factor in enhancing operational efficiency and harvesting rates. Various effective separation strategies are available, such as bending, lifting, rotation, pulling, or a combination of these methods [1]. Fan et al. employed finite element simulation techniques and conducted practical tests, revealing that a combined picking pattern involving both rotation and traction proved to be the most effective method for fruit separation [8]. In the present research, rotation was first selected as the separation technique, then a pull mode. The DC gearmotor starts rotating as soon as the apple is securely adsorbable, as shown in Fig. 4. The apple rotates as a result of the gears and transmission driving gears interacting. This circular motion helps to bend the stem of the fruit, which makes the pulling-off operation easier.

Fig. 3 Assembly diagram of end-effector

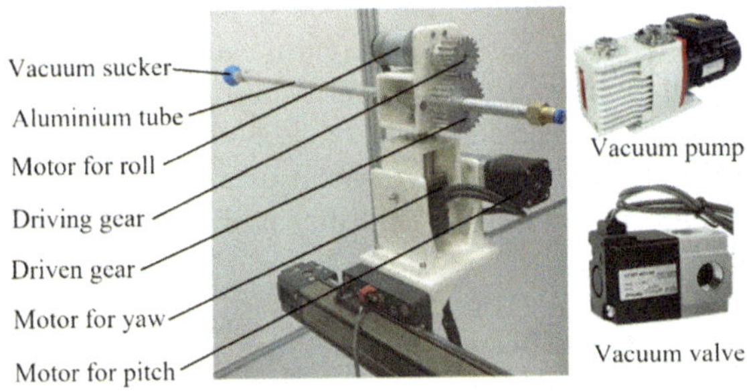

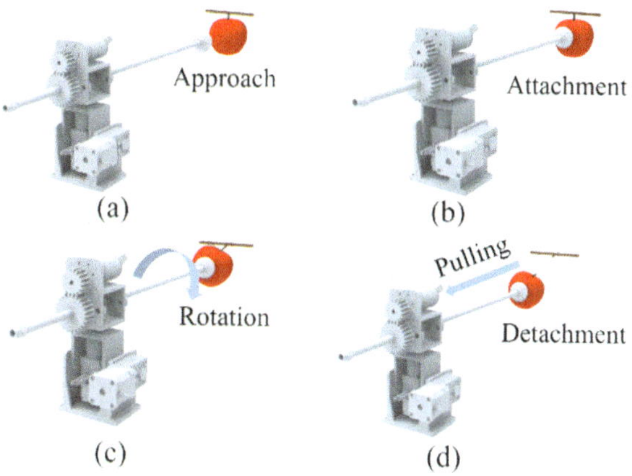

Fig. 4 Schematic diagram of end-effector separating apple: **a** approaching apple, **b** attachment apples, **c** rotation apple, and **d** pulling out and apple stem detachment

3.2.2 Integrated Harvesting Manipulator System

During manipulation, changes in pose significantly impact the spatial requirements of the system [32]. Each joint and link of the manipulator plays a crucial role in achieving a specific pose necessary for reaching any target position and orientation. A 3-DoF Cartesian system was developed and put into use in an attempt to efficiently place the 3R Degree of Freedom (DoF) end-effector close to the apple and enable rapid and accurate obtaining of the desired fruit location. This design choice minimizes errors and possesses the characteristic of minimizing attitude changes during manipulation. The integrated harvesting manipulator system, which includes the end-effector and Cartesian manipulator system, is depicted in Fig. 5. This system comprises six joints arranged in the PPPRRR (Prismatic, Prismatic, Prismatic, Revolute, Revolute, Revolute) configuration. Specifically, three prismatic joints govern the movement of the bottom Cartesian system, while three rotary joints control the end-effector's pitch, yaw, and roll. The sequence of prismatic joint placement is designated as follows: joint L1 moves along the x-axis, joint L2 along the z-axis, and joint L3 along the y-axis in the end-effector coordinates.

The three prismatic joints employ synchronous belt slide modules. The x-axis and z-axis slide modules have a lead of 950 mm and are driven by 86-stepper motors (Fanggong Intelligent Co., Ltd., China) with a holding torque of 4.5 Nm each. Additionally, the y-axis slide module has a lead of 750 mm and is driven by a 57-stepper motor (Fanggong Intelligent Co., Ltd., China) with a holding torque of 2.8 Nm. The pitch joint is powered by a stepper worm gear reducer motor (Type: 42–48; 1:100, Ming Yang Motor Co., Ltd., China) boasting a holding torque of 52 Nm. The yaw joint is driven by a stepper planetary gear reducer motor (Type: 42–40; 1:50, Ming Yang Motor Co., Ltd., China) with a holding torque of 20 Nm. Finally, the roll joint is actuated by a D.C. gearhead motor (Type: JGB37-545; 1:30, Ming Yang Motor Co., Ltd., China) featuring a holding torque of 12 Nm.

The manipulator developed for this investigation has several noteworthy benefits. It demonstrates three key features: (1) a wide operating range; (2) a small and simple

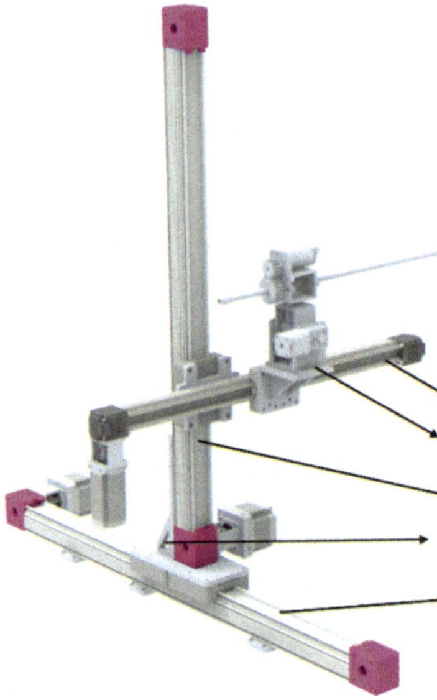

Fig. 5 Schematic structure of apple picking manipulator

construction; and (3) flexible motion performance. It offers a good number of degrees of freedom. The manipulator is a good fit for jobs involving harvesting because of these features. Its mechanical construction design is reasonably priced and very versatile.

The parameters of the manipulator are characterized by the degrees of freedom, joint types, link lengths, link offsets, and joint rotations. Figure 6 delineates the coordinate system of the manipulator and establishes its kinematic model. Modified Denavit-Hartenberg (D-H) parameters are employed to describe the joint coordinate transformations, and these modified D-H parameters are detailed in Table 1. As depicted in Fig. 6, the manipulator utilizes a global coordinate system represented as $X_0Y_0Z_0$, while the base coordinate system of the end-effector is $X_2Y_2Z_2$, and the suction cup coordinate system is $X_6Y_6Z_6$.

The transformation matrix of the manipulator can be derived using the modified D-H parameters. The position $[xyz]^T \in R$ of the sucker in the global coordinates of the manipulator can be calculated using this transformation matrix. The forward kinematic formulation of the manipulator is expressed in the following equation.

$$x = d_4 - a_6 * \sin\theta_5 \quad (1)$$

$$y = d_3 - (d_5 + d_6)sin\theta_4 + a_6cos\theta_4cos\theta_5 \quad (2)$$

$$z = a_4 + (d_5 + d_6)cos\theta_4 + a_6sin\theta_4cos\theta_5 \quad (3)$$

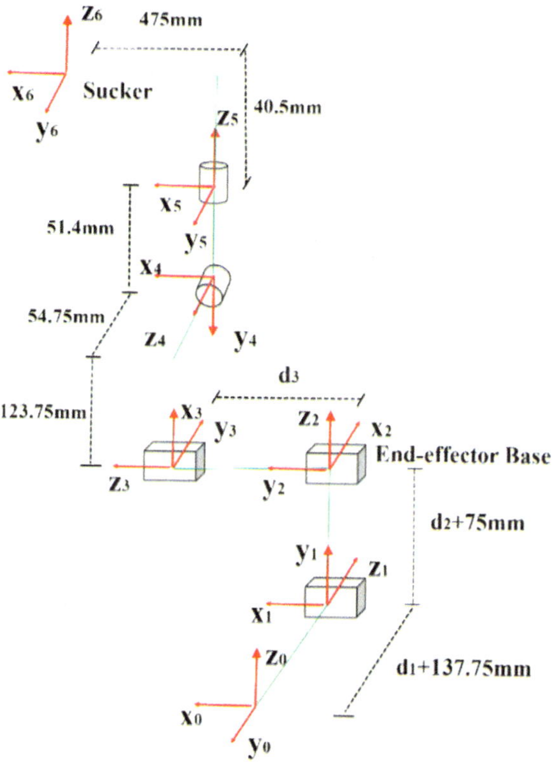

Fig. 6 Kinematic model of manipulator based on modified D-H parameter method

Table 1 Modified D-H parameters of the manipulator

Link	α_i/°	a_i/mm	d_i/mm	θ_i/°	Offset/°	Joint limit
Link1	90	0	d1 + 137.75	0	0	950 mm
Link2	– 90	0	d2 + 75	0	– 90	950 mm
Link3	– 90	0	d3	0	– 90	550 mm
Link4	– 90	123.75	54.75	θ_4	– 90	(– 25°, 25°)
Link5	– 90	0	51.4	θ_5	0	(– 25°, 25°)
Link6	0	475	40.5	0	0	0

The solution to the inverse kinematics equation for the manipulator is obtained through the forward kinematics equation, as depicted in the following equation.

$$d_3 = y - \sqrt{(d_5 + d_6)^2 + (a_6^2 - (d_4 - x)^2) - (z - a_4)^2} \quad (4)$$

$$\theta_4 = \arcsin\left(\frac{z - a_4}{\sqrt{(d_5 + d_6)^2 + (a_6^2 - (d_4 - x)^2)}}\right) - \varphi \tag{5}$$

$$\theta_5 = \arcsin\left(\frac{d_4 - x}{a_6}\right) \tag{6}$$

where

$$tan\varphi = \frac{d_5 + d_6}{\sqrt{a_6^2 - (d_4 - x)^2}} \tag{7}$$

3.2.3 Design of Visually Assisted Localization Systems

Apple detection involves segregating apples from the background, which includes elements such as leaves, branches, and trunks. Apple localization encompasses the computation of the 3D spatial position of the detected apples in relation to the camera coordinate system. Intel RealSense D435i RGB-D camera was employed to capture environmental information due to its affordability and accuracy, offering a resolution of 1280 × 720 for active stereo depth.

The detection and localization of apples in intricate orchard environments pose challenges, including partial shading by leaves and branches, color variations due to diverse light conditions, and varying ripeness levels. The YOLOv5 model stands out for its real-time capability, reliability, and robustness [20]. In addressing these challenges, YOLOv5 was employed for apple detection.

After detecting the apple using YOLOv5, the RGB and depth images are aligned. Depth information is then extracted from the depth map using the center point position of the 2D box. This operation finally determines the apple's spatial location, as seen in Fig. 7.

Upon acquiring the coordinates of the apple under the camera, the next step involves obtaining the transformation matrix between the camera and the manipulator through hand-eye calibration. Subsequently, the coordinates of the apple under the manipulator are solved. Depending on where the vision sensor is mounted on the manipulator, there are two mounting methods: eye-in-hand and eye-to-hand. For eye-to-hand, the camera is located globally and has a wide field of view, but the camera's depth information can be highly inaccurate; for eye-in-hand, the camera is located near the end-effector and the depth information is more accurate, but limits the camera's field of view. In this study, the camera was horizontally mounted on a frame at the end of the y-axis of the manipulator, presenting an eye-in-hand pattern for the x- and y-axes. Conversely, an eye-out-of-hand pattern was implemented for the z-axis and end-effector. The apple harvesting manipulator was subjected to hand-eye calibration, following the guidelines provided by Hu et al. [17]. Table 2 lists the rotation and translation matrices used in this process.

Fig. 7 Apple detection and localization based on yolov5

Table 2 Rotation and translation matrix for hand-eye calibration

Rotation matrix			Translation matrix
0.9998	– 0.0068	0.0136	22.7559
0.0128	0.0313	– 0.9994	64.9187
0.0065	0.9995	0.0314	– 542.9123

3.2.4 Control System Design

A graphical user interface (GUI) was created in PyCharm to facilitate the control of the manipulator system. Figure 8 provides a detailed overview of the control system. The central laptop served as the communication host, establishing connections with the RGB-D camera, the motion control card (Fangong Intelligent Co., Ltd., China), and the microcontroller (Arduino Mega 2560, Arduino Inc.) via USB, Ethernet, and USB cables, respectively. The motion control card is linked to three stepper motor drivers to regulate the 3-DoF Cartesian system using PWM, while the Arduino is connected to two stepper motor drivers and a D.C. motor driver to manage the end-effector through PWM. The vacuum pump and vacuum valve are initially configured in the 'open' position during regular operation, enabling the sucker to attract apples as it approaches the target position. The relay module is employed to close the vacuum valve, effectively controlling the cessation of the vacuum supply. The relay connection time is set to 0.75 s, ensuring proper placement of the apple.

The comprehensive procedure of the robot can be outlined as follows. Commencing the harvesting experiment involves initially positioning the cart at a predetermined distance in front of the fruit tree. Subsequently, the manipulator, consisting of the Cartesian system and the end-effector, is configured to a predefined initial position to execute the ensuing movements. The RGB-D camera takes pictures and uses 3D localization and fruit recognition to determine whether or not there are

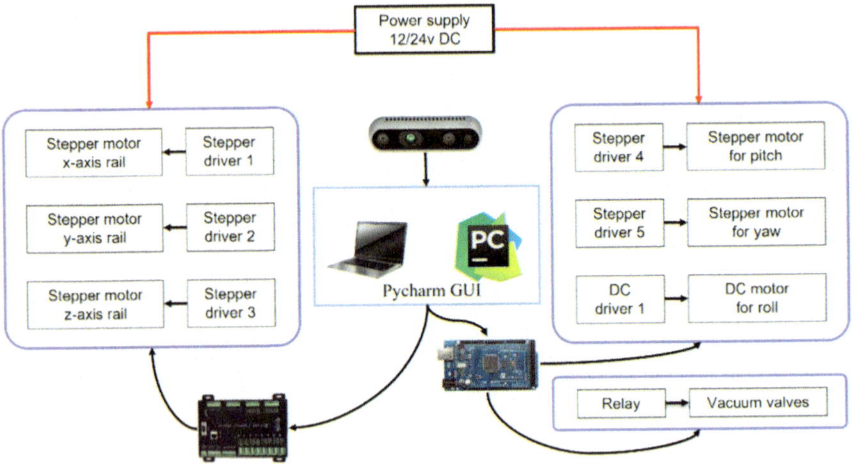

Fig. 8 Schematic diagram of the control system

apples that are suitable for picking within its range of vision. If no apples are detected, the x-axis and z-axis of the Cartesian system are simultaneously adjusted until apples come into view. The hand-eye matrix and kinematic inverse solution move the y-axis and end-effector closer to the apple. The apple is successfully adsorbed; at this point, the y-axis retracts to release it from the D.C. motor, and the motor rotates in a circular motion. When the apples reach their destination, the vacuum solenoid is closed by activating the relay, making arranging them easier. Subsequently, the system reevaluates the presence of pickable apples, and if detected, the separation process is reiterated until there are no pickable apples within the camera's range. Following this, an assessment is made to determine if the Cartesian system remains within its working range. The x- and z-axes are adjusted for the recognition process if within range. The vehicle is moved to the next tree if the Cartesian system reaches a predetermined limit, as seen in Fig. 9.

3.3 Harvesting Experiments with Manipulator

The prototype of the harvesting manipulator integrates an RGB-D vision sensor, a three-degree-of-freedom Cartesian system, an end-effector, a control system, and a mobile vehicle. The parameters of the robot prototype were evaluated through picking experiments conducted in a field orchard.

From October 23 to October 31, 2023, a field experiment was conducted on 'Fuji' apples at Beijing Shengshi Fruit King Agricultural Science and Technology Development Limited Liability Company (latitude 40°21′N, longitude 116°43′E). The experimental trees were approximately 5 years old. The apples were categorized based on the equatorial diameter (D) of the fruit, with $D \geq 80$ mm classified as large,

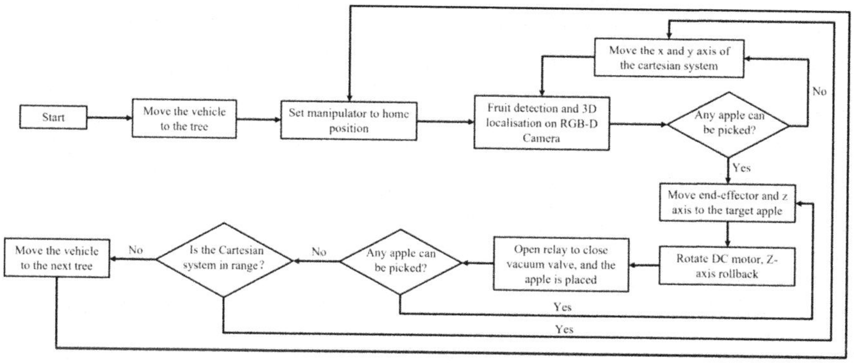

Fig. 9 Adsorption picking control flowchart

70 mm ≤ D < 80 mm as medium, and 65 mm ≤ D < 70 mm as small. Fruits with D < 65 mm were considered secondary and were not included in the data statistics.

Experiment 1 aimed to establish the size of the sucker and the effective picking method. Experiment 2 focused on determining the critical negative pressure value needed for successful apple separation, while Experiment 3 sought to identify the ideal picking adsorption position and its permissible error. During the field test, the distance between the side of the mobile vehicle and the test tree row was maintained between 0.6 and 0.8 m.

3.3.1 Experiments to Establish Sucker Size and Picking Method

The suction force generated under the same negative pressure increases with the size of the suction cup; however, the variable shape of the apple poses a challenge, as a huge size may hinder the establishment of an optimal adsorption relationship. Consequently, the shape parameters of the fruit are crucial for determining the appropriate suction cup size. To guarantee the reliable adsorption of vacuum suction cups, it is imperative to maintain a specific distance between the adsorption area and the edge of the apple. Utilizing shape of the fruit, two vacuum suckers with diameters of 33 and 43 mm were initially chosen. Sucker was selected through experimentation with three types of apples: large, medium, and small. 30 separation experiments were conducted for each of the rotations followed by pull mode and the pull mode to validate the reliability of the rotation followed by pull mode.

3.3.2 Negative Pressure Threshold Establishment Experiment

In this study, vacuum adsorption technology was employed as the principle for apple separation, focusing on determining its critical negative pressure value during apple adsorption. The negative pressure threshold represents the minimum level required for effective picking and serves as a crucial parameter to ensure the manipulator can reliably adsorb apples while minimizing power consumption.

To systematically investigate the impact of apple adsorption, a series of picking experiments were conducted on fruits of varying sizes: large, medium, and small. The objective was to explore the critical negative pressure required for separating apples within a range of air pressures from − 55 to − 75 kPa, adjusting the negative pressure values gradually in 5 kPa increments through a vacuum pressure gauge.

3.3.3 Experiment to Establish Adsorption Position and Permissible Error

Vacuum adsorption offers the advantage of accommodating a certain degree of localization error. However, there is a lack of studies focusing on the vacuum adsorption point and the permissible amount of error. This experiment aims to investigate the optimal adsorption point for vacuum adsorption of apples and determine the allowable error.

Presently, three primary methods are commonly employed to obtain 2D information about apples. The first method involves directly selecting the center of the image as the localization point. The second method utilizes the minimum outer circle of the fruit contour, determining the center point through the circle's center and radius [19, 29]. The third method involves locating the center using the centroid [10]. The selection of the center of the image and the minimum outer circle is straightforward to determine, providing an intuitive choice for the adsorption position. However, for irregularly shaped apples, these methods may not accurately identify the true center of the fruit, leading to imprecise picking positions.

On the other hand, the centroid considers the center of gravity of the shape and proves more accurate for irregularly shaped fruit. However, when the calyx of the fruit is slightly outward-facing, the centroid shifts downward, compromising its accuracy. In this study, the equatorial center is regarded as a key point in the circle-like shape, potentially serving effectively in establishing the apple adsorption point.

At a negative pressure of − 75 kPa, utilizing the image center as the reference point, the apple was approached and absorbed with a random deviation, and the deviation was recorded for the four specified points. The specific process involved the following steps: first, introducing a random positional deviation, activating the air pump to adsorb the apple, and marking the position at the suction cup; subsequently, deactivating the air pump to release the apple and capturing the current image. Following this, the manipulator is controlled to the marked suction cup position, the air pump is reactivated to adsorb the apple, and the picking action is executed. In case of separation failure, the operation was repeated. Considering that unsuccessful picking attempts could cause damage to the fruit stalk connection, three consecutive failures were recorded as an overall failure.

4 Results and Discussion

4.1 Size of Sucker and Picking Method Established

Effective adsorption on apples of all sizes and orientations is demonstrated by the 33 mm suction cup, with minimal requirements on the contact surface. Conversely, medium to large apples can be securely held by the 43 mm suction cup, but for smaller fruits, a perfect contact area is necessitated. In such instances, insecure adsorption, and the potential for air leakage, may result from even a slight error.

In our 30 picking experiments using the separation method, 19 apples were successfully picked in the pull mode, and the separation flow is illustrated in Fig. 10. Additionally, 25 apples were successfully picked using the rotation followed by the pull mode, and the corresponding separation flow is depicted in Fig. 11. The experiments confirmed the greater reliability of the rotation followed by pull mode. Consequently, it was ultimately adopted as the preferred method for apple separation.

(a) (b)

Fig. 10 Actual process of the pull mode: **a** adsorption of apples, and **b** pulling back to separate apples

(a) (b) (c)

Fig. 11 Actual process of the rotation followed by pull mode: **a** adsorption of apples, **b** rotating apples, and **c** pulling back to separate apples

4.2 Negative Pressure Critical Value Experimental Results

In the initial picking experiments conducted at – 55 kPa air pressure, a total of 45 apples were tested, resulting in the successful picking of 20 apples, yielding an overall success rate of 44.4%. At – 60 kPa air pressure, 22 apples were successfully picked, maintaining a comparable overall success rate of 48.8% with no significant change. Subsequently, at – 65 kPa, the success rate notably increased to 84.4%, demonstrating a substantial surge of 33.4%. To ensure the robustness of the findings, the sample size was expanded. For the picking experiments at – 70 kPa, a total of 60 apples were tested, with 49 were successfully picked, resulting in an overall success rate of 81.7%. Finally, in the experiments at – 75 kPa, 50 apples were successfully picked, achieving an overall success rate of 83.3%. Detailed results are presented in Table 3.

The findings suggest that an air pressure of – 65 kPa appears to be the critical value for achieving accurate apple separation. To eliminate the possibility of chance, additional picking experiments were conducted at the same air pressure of – 65 kPa, involving a total of 60 apples. Among them, 49 were successfully picked, resulting in an overall success rate of 81.7%. The outcomes of this series of experiments clearly demonstrated a significant increase in the success rate of apple picking under the air pressure of – 65 kPa, establishing it as the critical threshold for negative pressure picking. These results serve as a crucial reference for future research in this domain.

In the field test, there was no significant improvement in the success rate despite increasing the negative pressure to – 75 kPa. To investigate the underlying reasons, a careful examination of the failure cases from the last three sets of experiments was conducted, and the factors contributing to the separation failures were analyzed, as depicted in Table 4. The primary observations are as follows: unusual growth postures of the target fruit, such as excessive outward growth of the calyx, leading to a reduction in tensile force due to the large contact angle at the end of the actuator; target fruits situated on slender branches, where branch movement with the apple during picking fails to generate effective separation force; obstruction caused by the target fruit being in front of a branch, preventing adsorption or resisting movement due to branch action, resulting in failure; and individual fruits being immature, requiring

Table 3 Statistics of harvesting under different negative pressures

Negative pressure (kpa)	Large fruit ((D ≥ 80 mm) (success/overall)	Medium fruit (70 mm ≤ D<80 mm) (success/overall)	Small fruit (65 mm ≤ D<70 mm) (success/overall)	Average
– 55	5/8	10/22	5/15	20/45
– 60	6/7	13/26	3/12	22/45
– 65	14/16	16/19	7/10	37/45
– 70	12/13	24/30	13/17	49/60
– 75	18/20	25/29	7/11	50/60

Table 4 Failure statistics at different negative pressures

Case of failure	Under – 65 kpa	Under – 70 kpa	Under – 75 kpa
Calyx excessively outward	3	4	3
Slender branch	5	3	2
Shade from a tree branch	2	2	2
Immature	1	2	3

excessive force at the peduncle-branch junction, causing the main branch to move with the apple, creating resistance that leads to detachment from the adsorption.

4.3 Establishment of Adsorption Position and Permissible Error

Overall, a total of 100 sets of experiments were conducted, resulting in the successful picking of 82 apples. The distribution of adsorption locations was analyzed using the four theoretical picking points mentioned previously as benchmarks.

The process for establishing the adsorption location is depicted in Fig. 12. The fruit contour within the apple Region of Interest (ROI) was extracted using Anylabeling (Anylabeling is obtained from the open source website GitHub: https://github.com/vietanhdev/anylabeling). Subsequently, the image center, minimum outer circle center (MOCC), centroid, and the equatorial center of the apple were extracted as theoretical adsorption points through image processing. The actual adsorption point was determined by extracting the adsorption circle through threshold segmentation and Hough detection, using the circle's center as the actual adsorption point. The distance from the actual adsorption point to each theoretical adsorption point for every apple was calculated as a percentage of the projection on the vectors of the two sides of the minimum outer rectangle of the apple. The mean and variance of these distances were then computed.

The mean and variance of these distances were then computed, with the results presented in Table 5. The equatorial center exhibited the smallest mean deviation and the most uniform distribution, making it the selected theoretical adsorption point. A scatter plot was generated to visualize the deviation metric, as shown in Fig. 13, revealing that allowable errors are concentrated in the quadrilateral region.

4.4 Damage and Fruit Stem Retention

No significant damage was observed in the apples after picking. Subsequently, twenty-five apples were selected for further observation after being stored at room temperature in the laboratory for 15 days. Observations were recorded on the 7th

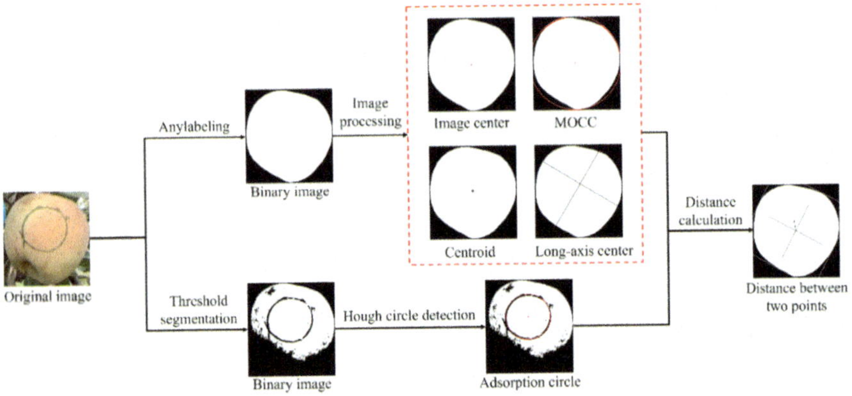

Fig. 12 Deviation measurement flowchart

Table 5 Mean and variance of deviation for each theoretical picking point

Theoretical picking point	Average	Variance
Image center	9.87	26.24
MOCC	9.91	24.56
Centroid	9.93	24.01
Equatorial center	9.62	18.19

Fig. 13 Scatter plot of the center of the equatorial center

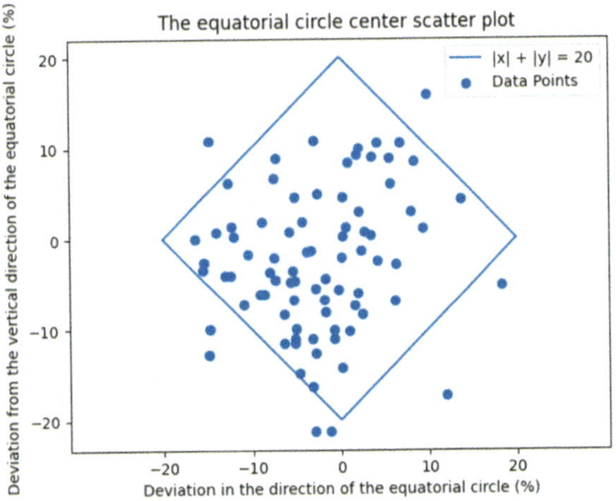

and 15th days. After seven days of storage, there was no discernible change in the overall appearance of the apples or the suction cup attachment area. After 15 days of storage, some surface wrinkling was noted on the apples, but the suction cup area still exhibited no signs of bruising or damage. This suggests that the harvesting method

Table 6 Comparison with other studies

Grabbing method	Separation method	DoF	Damaged	Power	Flexible	Reference
Suck into tube	Inhalation	–	Yes	High	Yes	Burwood-Taylor [3]
Rigid grip	Twist off	7	Yes	Low	Yes	Silwal et al. [26]
Soft grip	Pull off	7	Yes	Low	Yes	Hohimer et al. [16]
Vacuum attachment	Twist off	3	No	High	No	Zhang et al. [34]
Vacuum attachment	Rotation – pulling	4	No	Low	No	Hu et al. [17]
Vacuum attachment	Rotation + pulling	6	No	Low	Yes	Our study

employed in this study had a positive impact on the overall quality of the apples after harvesting.

Throughout the entire experiment, a total of 445 attempts were made, resulting in the successful picking of 334 apples. Impressively, 82.9% of the picked apples retained their fruit stems. The preservation of fruit stems played a vital role in maintaining the freshness and texture of the apples.

4.5 Comparison with Existing Studies

The newly developed manipulator demonstrated a simple structure, low power consumption, a large stroke, and ease of canopy penetration when compared to the system described by Burwood-Taylor [3]. In contrast to the work by Silwal et al. [26] and Hohimer et al. [16], this manipulator avoids causing damage to the target fruits and doesn't collide with neighboring fruits during the picking process. Compared to the studies by Zhang et al. [34] and Hu et al. [17], this 6-DoF manipulator shows greater flexibility, allowing a wide range of picking, as shown in Table 6.

5 Conclusions

This study presents the development of a vacuum-based 6-DoF non-destructive apple harvesting manipulator. The prototype integrates an RGB-D camera, a 3-DoF Cartesian system, and a 3R-DoF end-effector for efficient harvesting. In harvesting experiments, 33 mm suction cups demonstrated superior adsorption compared to 43 mm suction cups, and the rotation followed by the pull mode proved more reliable than the pull mode. Negative pressure of − 65 kPa emerged as the critical value for effective apple separation, with an 81.7% success rate, and the equatorial center was identified as the optimal theoretical point for adsorption with an allowable two-dimensional error mean value of 9.8%. The negative pressure adsorption picking method ensures minimal damage, with a fruiting stem abscission rate of 17%. While the developed prototype accomplishes the task of apple picking, manual assistance is currently required for fruit positioning.

Future work aims to integrate a far-near vision system and posture estimation with additional cameras for fully autonomous harvesting, enhance the hardware and software structure of the manipulator, and improve harvesting efficiency by incorporating multiple end-effectors to cover the entire fruit tree.

Acknowledgements This research is supported by the National Key Research and Development Program of China (2023YFE0122600). The experiment was conducted at Beijing Shengshi Fruit King Agricultural Technology Co., LTD.

References

1. Bac CW, Hemming J, van Henten EJ (2014) Stem localization of sweet-pepper plants using the support wire as a visual cue. Comput Electron Agric 105:111–120. https://doi.org/10.1016/j.compag.2014.04.011
2. Bu LX, Chen CK, Hu GR, Sugirbay A, Chen J (2020) Technological development of robotic apple harvesters: a review. Inmateh Agric Eng 61(2). https://doi.org/10.35633/inmateh-61-17
3. Burwood-Taylor L (2017) Abundant Robotics raises $10m series A for apple picking robot led by GV. https://agfundernews.com/abundant-robotics-raises-10m-series-a-for-apple-picking-robot
4. Ceccarelli M, Criado EJ, Figliolini G, Mata AS, Ottaviano E (2000) Designing a robotic gripper for harvesting horticulture products. Robotica 18(1):105–111. https://doi.org/10.1017/S026357479900226X
5. Chen Q, Yin CK, Guo ZL, Wang JP, Zhou HP, Jiang XS (2023) Current status and future development of the key technologies for apple picking robots. Trans Chin Soc Agric Eng 38(4):1–15. https://doi.org/10.11975/j.issn.1002-6819.202209041
6. De Kleine ME, Karkee M (2015) A semi-automated harvesting prototype for shaking fruit tree limbs. Trans ASABE 58(6):1461–1470. https://doi.org/10.13031/trans.58.11011
7. Elfferich JF, Dodou D, Santina CD (2022) Soft robotic grippers for crop handling or harvesting: a review. IEEE Access 10:75428–75443. https://doi.org/10.1109/ACCESS.2022.3190863
8. Fan P, Yan B, Wang M, Lei X, Liu Z, Yang F (2021) Three-finger grasp planning and experimental analysis of picking patterns for robotic apple harvesting. Comput Electron Agric 188:106353. https://doi.org/10.1016/j.compag.2021.106353
9. Fathallah FA (2010) Musculoskeletal disorders in labor-intensive agriculture. Appl Ergon 41(6):738–743. https://doi.org/10.1016/j.apergo.2010.03.003
10. Feng J, Wang S, Liu G, Zeng L (2011) A separating method of adjacent apples based on machine vision and chain code information. IFIP Adv Inf Commun Technol 368:258–267
11. Fountas S, Mylonas N, Malounas I, Rodias E, Hellmann Santos C, Pekkeriet E (2020) Agricultural robotics for field operations. Sensors 20(9):2672. https://doi.org/10.3390/s20092672
12. Gongal A, Amatya S, Karkee M, Zhang Q, Lewis K (2015) Sensors and systems for fruit detection and localization: a review. Comput Electron Agric 116:8–19. https://doi.org/10.1016/j.compag.2015.05.021

13. Grandd A, Esnon A (1984) Robotic harvesting of apples. ASAE Publ 4–84:112–113
14. He L, Fu H, Sun D, Karkee M, Zhang Q (2017) Shake-and-catch harvesting for fresh market apples in trellis-trained trees. Trans ASABE 60(2):353–360. https://doi.org/10.13031/trans.12067
15. He L, Zhang X, Ye Y, Karkee M, Zhang Q (2019) Effect of shaking location and duration on mechanical harvesting of fresh market apples. Appl Eng Agric 35(2):175–183. https://doi.org/10.13031/aea.12974
16. Hohimer CJ, Wang H, Bhusal S, Miller J, Mo C, Karkee M (2019) Design and field evaluation of a robotic apple harvesting system with a 3d-printed soft-robotic end-effector. Trans ASABE 62(2):405–414. https://doi.org/10.13031/trans.12986
17. Hu G, Chen C, Chen J, Sun L, Sugirbay A, Chen Y, Jin H, Zhang S, Bu L (2022) Simplified 4-dof manipulator for rapid robotic apple harvesting. Comput Electron Agric 199:107177. https://doi.org/10.1016/j.compag.2022.107177
18. Jia W, Zhang Y, Lian J, Zheng Y, Zhao D, Li C (2020) Apple harvesting robot under information technology: a review. Int J Adv Robot Syst 17(3):255688461. https://doi.org/10.1177/1729881420925310
19. Jiao Y, Luo R, Li Q, Deng X, Yin X, Ruan C, Jia W (2020) Detection and localization of overlapped fruits application in an apple harvesting robot. Electronics 9(6):1023. https://doi.org/10.3390/electronics9061023
20. Liu X, Zhang Z, Igathinathane C, Flores P, Zhang M, Li H, Han X, Ha T, Ampatzidis Y, Kim H (2024) Infield corn kernel detection using image processing, machine learning, and deep learning methodologies under natural lighting. Expert Syst Appl 238:122278. https://doi.org/10.1016/j.eswa.2023.122278
21. Luo T, Escalante CL (2017) Us farm workers: what drives their job retention and work time allocation decisions? Econ Labour Relat Rev 28(2):270–293. https://doi.org/10.1177/1035304617703933
22. Monta M (2000) Robots for bioproduction systems. IFAC Proc Vol 33(29):1–11. https://doi.org/10.1016/S1474-6670(17)36744-7
23. Peterson DL, Miller SS (1989) Advances in mechanical harvesting of fresh market quality apples. J Agric Eng Res 42(1):43–50. https://doi.org/10.1016/0021-8634(89)90038-3
24. Peterson DL, Wolford SD (2003) Fresh–market quality tree fruit harvester part II: apples. Appl Eng Agric 19(5):545. https://doi.org/10.13031/2013.15314
25. Sarig Y (1993) Robotics of fruit harvesting: a state-of-the-art review. J Agric Eng Res 54(4):265–280. https://doi.org/10.1006/jaer.1993.1020
26. Silwal A, Davidson JR, Karkee M, Mo C, Zhang Q, Lewis K (2017) Design, integration, and field evaluation of a robotic apple harvester. J Field Robot 34(6):1140–1159. https://doi.org/10.1002/rob.21715
27. Subeesh A, Mehta CR (2021) Automation and digitization of agriculture using artificial intelligence and internet of things. Artif Intell Agric 5:278–291. https://doi.org/10.1016/j.aiia.2021.11.004
28. Wang X, Kang H, Zhou H, Au W, Wang MY, Chen C (2023) Development and evaluation of a robust soft robotic gripper for apple harvesting. Comput Electron Agric 204:107552. https://doi.org/10.1016/j.compag.2022.107552
29. Wang D, Song H, Tie Z, Zhang W, He D (2016) Recognition and localization of occluded apples using k-means clustering algorithm and convex hull theory: a comparison. Multimed Tools Appl 75(6):3177–3198. https://doi.org/10.1007/s11042-014-2429-9
30. Xu J, Gu B, Tian G (2022) Review of agricultural IoT technology. Artif Intell Agric 6:10–22. https://doi.org/10.1016/j.aiia.2022.01.001
31. Yuan J (2020) Research progress analysis of robotics selective harvesting technologies. Trans Chin Soc Agric Machi 51:1–17. https://doi.org/10.6041/j.issn.1000-1298.2020.09.001
32. Zahid A, Mahmud MS, He L, Choi D, Heinemann P, Schupp J (2020) Development of an integrated 3r end-effector with a Cartesian manipulator for pruning apple trees. Comput Electron Agric 179:105837. https://doi.org/10.1016/j.compag.2020.105837

33. Zhang Z, Igathinathane C, Li J, Cen H, Lu Y, Flores P (2020) Technology progress in mechanical harvest of fresh market apples. Comput Electron Agric 175:105606. https://doi.org/10.1016/j.compag.2020.105606
34. Zhang K, Lammers K, Chu P, Li Z, Lu R (2021) System design and control of an apple harvesting robot. Mechatronics 79:102644. https://doi.org/10.1016/j.mechatronics.2021.102644
35. Zhang Z, Heinemann PH, Liu J, Schupp JR, Baugher TA (2016) Design and field test of a low-cost apple harvest-assist unit. Trans ASABE 59(5):1149–1156. https://doi.org/10.13031/trans.59.11708
36. Zhang Z, Heinemann PH (2017) Economic analysis of a low-cost apple harvest-assist unit. Horttechnology 27(2):240–247. https://doi.org/10.21273/HORTTECH03548-16
37. Zheng W, Xie Y, Zhang B, Zhou J, Zhang J (2021) Dexterous robotic grasping of delicate fruits aided with a multi-sensory e-glove and manual grasping analysis for damage-free manipulation. Comput Electron Agric 190:106472. https://doi.org/10.1016/j.compag.2021.106472
38. Zhou H, Wang X, Au W, Kang H, Chen C (2022) Intelligent robots for fruit harvesting: recent developments and future challenges. Precis Agric 23(5):1856–1907. https://doi.org/10.1007/s11119-022-09913-3

Research Progress of Assisted Pollination Technology

Qu Zhang, Zhao Zhang, Xu Li, and Liling Yang

Abstract To systematically understand the importance of pollination in both natural ecosystems and agricultural production, this paper reviews the development and current state of flower pollination technologies, analyzes the advantages and disadvantages of existing mechanical pollination methods, and explores future directions for pollination technologies. The review encompasses research on assisted pollination globally and domestically, summarizing mainstream research approaches, methodologies, and findings. It also anticipates future research trends, focusing on factors such as pollination efficiency and fruit setting rates compared between natural, artificial, and mechanical pollination methods. The literature review identifies three primary types of assisted pollination methods: artificial assisted pollination, robot pollination, and drone pollination. Artificial assisted pollination, while beneficial for enhancing orchard yields, is labor-intensive and time-consuming. In contrast, robot and drone pollination technologies offer higher autonomy but require specific environmental conditions and currently exhibit lower fruit setting rates. Looking ahead, the integration of mechanical pollination with advanced agricultural technologies holds promise for developing precise pollination techniques tailored to different plant species and growth environments. This approach aims to improve overall pollination efficiency and fruit quality in agricultural production systems.

Keywords Pollination · Robotic · Drone

Q. Zhang · Z. Zhang (✉)
Key Laboratory of Smart Agriculture System Integration, Ministry of Education, China Agricultural University, Beijing 100083, China
e-mail: zhaozhangcau@cau.edu.cn

Key Laboratory of Agricultural Information Acquisition Technology, Ministry of Agriculture and Rural Affairs, China Agricultural University, Beijing 100083, China

X. Li
College of Information Engineering, Tarim University, Aral 843300, Xinjiang, China

L. Yang
Institute of Agricultural Mechanization, Xinjiang Academy of Agricultural Sciences, Ürümqi 830091, Xinjiang, China

Y. Liu et al. (eds.), *New Technologies Applied in Apple Production*, Smart Agriculture 10, https://doi.org/10.1007/978-981-97-7778-5_3

1 Introduction

Pollination refers to the process of transferring plant pollen from anther to stigma, which is a necessary process for plants to form fruits. Of the approximately 352,000 plant species on Earth, nearly 90% rely on insects or other animals for pollination and reproduction, and 85% of flowering plants rely on bees for pollination. According to research, bee pollination can greatly improve the yield and quality of crops, such as increasing the yield of rice by 2.5–7.1%; Cotton production increased by 38%, lint length increased by 8.6%; Rapeseed yield increased by 18.7 ~ 37%, oil yield increased by 10%; A 22% increase in apple production, for example, could increase the economic value of world agriculture by more than 100 billion euros per year, and the annual output value of global crops directly affected by pollinators is between 235 and 577 billion US dollars. Therefore, pollination is crucial for maintaining ecosystem balance and agricultural production, and has important economic significance. However, with the abuse of pesticides, environmental pollution and destruction, and the reduction of vegetation area, the number of bees is declining sharply. Bee pollination, as "aerial agriculture", has become a specialized industry in Western countries and is an important measure to increase agricultural production. The United States, for example, rents 1 million honeybees a year to pollinate crops, increasing crop production by about $20 billion. According to the Food and Agriculture Organization of the United Nations (FAO), the number of managed beehives worldwide has increased by 45% over the past half century, while the portion of agriculture dependent on animal pollination has increased by 300%, increasing the demand for agricultural pollination services and putting pressure on global pollination capacity. In order to reduce the demand for pollination on natural conditions, currently the pollination of orchards and greenhouses is mainly completed manually. This method requires operators to pollinate one by one, with high pollination accuracy, but it is difficult for humans to contact the flowers at the top of the branches, which may easily lead to the staff's waist and elbow injuries. Therefore, this is an intensive work with high labor intensity, high labor cost and lack of digital accurate management. In order to save labor and time costs, improve pollination efficiency, and adapt to large-scale production, mechanical pollination research is a trend of current research in various countries, and some pollination devices have been commercialized. Examples include New Zealand company Pollen Plus [1] (https://www.pollenplus.co.nz/quadduster) kiwi pollinator "QuadDuster", US company Dropcopter's pollinator drone, Israeli company Arugga's greenhouse tomato pollinator robot and Edete's (https://edetepta.com/pollination-service/) almond tree pollinator. The scale and mechanization of pollination can be realized on the basis of ensuring the pollination fruit setting rate. However, these pollinators generally use spray pollination, and large-scale pollination equipment requires workers to operate and maintain, such as Edete's almond pollinators, which require operators to drive vehicles through almond forests. Fast, real-time, accurate and autonomous pollination is the bottleneck of current mechanical pollination technology. Precision pollination is also considered to be one of the key technologies for future precision agriculture [2].

In this review, we take greenhouse, orchard, field as the pollination background, pollination effect as the starting point, from the research Angle, method and content of the general overview. Firstly, the auxiliary pollination methods commonly used in greenhouses and orchards are discussed, and their advantages and disadvantages are analyzed. Secondly, the components and functions of the pollinator are described in detail. Since the crop morphology and topography of field crops are quite different from those of greenhouses and orchards, the pollination methods of field crops are reviewed in Sect. 3. Both manual and mechanical pollination methods are different from those described in Sect. 2. Finally, we discuss the future research trend of precise pollination of plants, and propose a method of precise pollination by drones in fruit gardens.

2 Universal Supplementary Pollination Methods

Both facility agriculture and orchards have a variety of pollination methods, such as bees, manual, mechanical, etc. Tomatoes and strawberries are common facility agriculture crops, tomatoes are self-pollinating crops, open planting can be pollinated by wind coal, greenhouse planting needs to be pollinated by insects. However, the tomato itself does not have nectaries, so bees do not pollinate it. In order to ensure the quality and fruit setting rate of the tomato, bumblebees are mainly used for pollination in production. Bumblebees have a slightly larger body, hard-working sex, humidity and low temperature tolerance, poor light source resolution, and good pollination effect on a variety of crops. The correct use of plant growth promoters can not only prevent falling flowers and fruit, improve fruit setting rate, but also accelerate fruit expansion, promote fruit ripening, and increase early yield, so the production is also generally used to dip flowers to improve tomato fruit setting rate and yield. Until now, many orchards have relied mainly on hired workers to manually dot the trees or spray pollen solutions or suspensions with hand-held or knapsack sprayers. Some orchards have also begun to use drone pollination, which is more efficient and more uniform than manual pollination. In the case of manual spraying, pollen is mainly concentrated below the canopy, while in the case of UAV spraying, pollen solution is mainly concentrated above the canopy. At present, there is no efficient and precise pollination method that has been widely applied.

This chapter reviews the common pollination methods in various scenarios such as facility agriculture and orchards, introduces bumble bee pollination and artificial pollination methods, and describes the development of robot pollination and its components and drone pollination in detail.

Fig. 1 Universal supplementary pollination methods: **a** Artificial assisted pollination method and **b** mechanical assisted pollination method

2.1 *Bumblebee Pollination*

In Fig. 1a-I, the process of bumblebee pollination begins with providing food for the bumblebees and relocating their hives into the greenhouse in the evening. The following morning, adjust the hive openings to allow bumblebees access to the flowers. It is crucial to avoid harmful pesticides inside the greenhouse. The optimal time for releasing bees is when approximately 5% of the tomato blossoms are open. Install insect nets on air vents to prevent bumblebees from escaping. Maintain temperatures between 8 and 32°C; activate the ventilation system during high temperatures and position hives on the back wall or ground during cooler periods.

2.2 *Artificial Pollination*

2.2.1 Dipping Flowers for Pollination

Tomatoes are typically self-pollinating plants, commonly assisted by hormone dipping methods to enhance production. The optimal time for hormone dipping is generally between 8 and 10 in the morning on sunny days. A small amount of 2, 4-dichlorophenoxyacetic acid (2, 4-D) liquid is applied with a brush lightly on the pedicel of just-opened female flowers. Care must be taken not to allow the medication to contact the stem and leaves to avoid plant damage. The best results are achieved when female flower petals are fully unfolded and trumpet-shaped. Adjusting the hormone concentration is crucial; for instance, Ma et al. [3] found that a concentration of 20 mg/kg of 2, 4-D was optimal for tomato flower dipping in early spring solar greenhouses. Higher concentrations are recommended during colder months (November to February) and lower concentrations during warmer months (March to

July). Excessive flower dipping can lead to labor, time, and medication costs, potentially causing issues like deformed, cracked, or hollow fruit. Residual hormones may also pose environmental concerns, impacting tomato quality and safety. Point pollination is another method used in orchards, involving the collection of pollen from the pollinating tree and its application using brushes, gauze balls, or paper sticks.

2.2.2 Vibrating Flowers for Pollination

Vibrator pollination involves both hand-held and knapsack methods where pollen is dispersed onto the style through vibration (Fig. 1a-III and a-IV). Operatives carry vibrating pollinators through inflorescences, causing pollen to oscillate and fall onto the stigma, facilitating natural pollination and fruit setting. The end effectors of pollinating robots in Sect. 2.3.3 below are often designed based on this principle, proving effective for self-pollinating plants. Both vibrator pollination and bumblebee pollination significantly enhance fruit setting rates, reduce deformities, improve fruit quality, and increase plant yields. However, ensuring optimal pollination timing for each flower can be challenging, and these methods can be time-consuming and labor-intensive. Care must be taken to avoid damaging inflorescences and plants. Sun compared the effectiveness of bumblebee pollination, hormone pollination, and vibrator pollination, concluding that bumblebee pollination is a safe, efficient, and cost-effective method, particularly suitable for tomato cultivation in solar greenhouses [4].

2.3 *Mobile Land Robot Pollination*

With the rapid development and integration of hardware and software, robotics technology has achieved significant breakthroughs in performance and functionality, leading to transformative changes across various sectors. Advanced algorithms and powerful hardware enable robots to perform increasingly complex and delicate tasks. In agriculture, robots have become indispensable, with the development and application of pollinator robots emerging as a major innovation focus. On February 23, 2024, we conducted a literature search in the Web of Science core database for the past 20 years (2003–2023) using the keyword "robotic pollination" The results indicated that the United States had the highest number of publications, followed by China, India, and New Zealand. However, the overall number of publications remains relatively small, reflecting limited research on robotic pollination in these countries (Fig. 2a). Since 2015, there has been an increase in research interest in this field, as evidenced by the rising number of publications per year. It is reasonable to anticipate that interest in robotic pollination will continue to grow with ongoing advancements in robotics and smart agriculture (Fig. 2c). Regarding publication types, the research outputs on robotic pollination technology are diverse. Articles account for the highest proportion (68%), followed by conference papers (38%), with relatively few review articles (6%). This diversity indicates sustained interest and exploration in robotic pollination technology. The numerous papers encompass basic theoretical and experimental research, as well as technological applications and effectiveness evaluations, providing both theoretical and practical support for technology development. Academic conferences serve as crucial platforms for scholars to share research findings. Through conference papers, researchers present the latest results, exchange technological advancements, and foster international cooperation. Although there are fewer review articles, they play a vital role in synthesizing research outcomes and guiding future research directions (Fig. 2d). To quickly grasp the core content of the literature, we created a CiteSpace visualization map (Fig. 2b) and analyzed the intermediary centrality of each keyword. The results show that the term "algorithm" has the highest centrality, underscoring the pivotal role of algorithm design and application in robotic pollination. Additionally, keywords such as "convolutional neural networks", "end effector", "flight", "machine vision", "autonomous pollination" and "artificial intelligence" also exhibit notable centrality. This indicates that robotic pollination technology involves interdisciplinary research. These keywords not only highlight the research direction of the field but also provide a multidimensional perspective for in-depth exploration of robotic pollination technology.

Greenhouse crops are generally low in height, uniformly distributed, and situated in simple terrains, making them ideal for robot operation. Autonomous and efficient greenhouse pollinator robots are considered the future of greenhouse crop

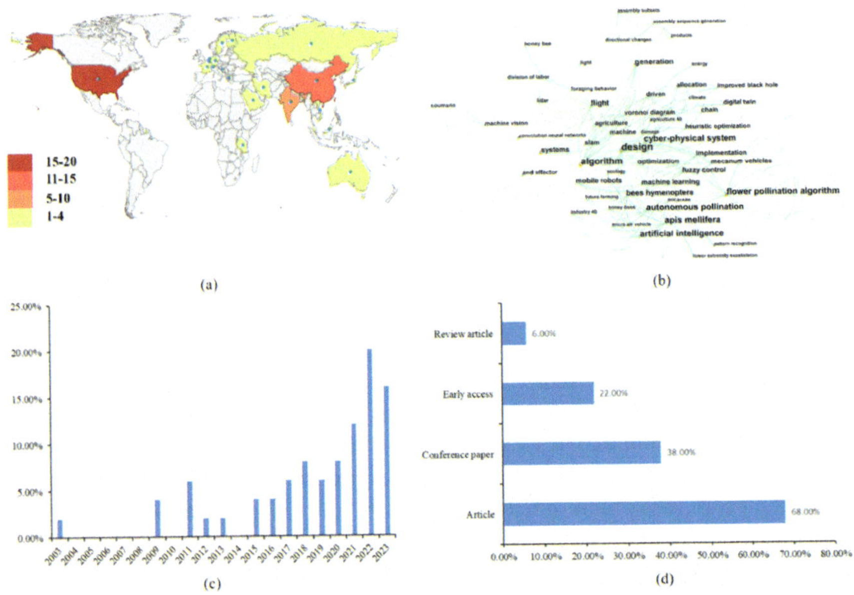

Fig. 2 Robotic pollination related publications: **a** Number of publications for individual countries, **b** CiteSpace map of different publication countries, **c** Number of publications per year and **d** Related areas to robotic pollination studies

pollination. Research on precise pollination in greenhouses has been conducted both domestically and internationally. For example, Yuan et al. [5] developed a hormone pollination robot for greenhouse tomatoes. This robot is capable of collision-free movement of the manipulator according to the planting conditions of tomatoes. It uses a binocular vision system to detect flowers based on their saturation, color, size, and other characteristics and determines the three-dimensional spray position of flowers using stereo vision. The average error between the detected distance and the measured injection position was found to be 6.4 mm. Internationally, the development of robotic pollination has progressed rapidly, resulting in more complete equipment. West Virginia University developed a ground-wheeled autonomous precision pollination robot named "BrambleBee" Utilizing synchronous positioning and mapping algorithms, LiDAR, and RGB-D cameras, this robot can create detailed geometric maps of its environment for route planning, autonomous navigation, flower recognition, positioning, and attitude detection. Pollen collection and pollination are conducted using an endoscope camera and a flexible end-effector (Fig. 1b-I) [6]. In 2019, Strader et al. [7] introduced an improved "BrambleBee" subsystem, enhancing flower recognition and end-effectors. This development made it the first precise and autonomous flower pollination system that integrates sensing, planning, control, and autonomous technologies to precisely interact with flowers, filling the technical gap of fully autonomous and precise pollination of single florets.

To maintain crop productivity, some companies are developing new pollination devices. New Zealand's Pollen Plus developed the "QuadDuster" equipped with mechanical pollen boxes, mechanical hairdryers, and sprayers for pollinating kiwi fruit orchards (Fig. 1b-II). The Israeli company Arugga built an AI robot for pollinating greenhouse tomatoes. This robot uses computer vision to identify flowers suitable for pollination and then sprays them with pulses of air to aid pollination. It has already successfully pollinated 10 hectares of tomatoes at the Australian Vegetable Costa Group, increasing tomato yield by about 20% without compromising quality (Fig. 1b-III). Mechanized pollination of date palm trees, walnuts, olives, almonds, and other tall trees began as early as the 1960s. The introduction of a piloted pollinator by the Israeli company Edete reduced labor requirements by 50–70% (Fig. 1b-IV) [8]. Gan-Mor et al. [9] also designed a stable mobile robot arm for precise spraying in date orchards. Robotic pollination involves a complex system design that integrates sensors, automation, mechanical design, machine vision, and other technologies. A complete pollination robot generally needs to meet the following criteria: (a) identify and locate the flowers in pollination in three-dimensional space, (b) approach and reach the target flower,and (c) accomplish pollination without damaging the flowers and plants.

2.3.1 Identification and Localization of Flowers

Identifying and locating flowers is the primary challenge for pollinating robots or drones. This task is complex due to factors such as occlusions, varying weather conditions, and lighting. Additionally, it is crucial for robots to determine whether flowers have already been pollinated, especially those without prior pollination records. With the maturation of image processing technology and the development of deep learning, there are increasingly sophisticated methods for flower recognition and positioning, leading to significant improvements in accuracy.

Boykov and Jolly [10, 11] provided and demonstrated a general interactive segmentation technique for N-dimensional images. Kim et al. [12] proposed a mobile flower recognition system based on differential image entropy (DIE) and contour features. Yuan et al. [5] developed a hormonal pollinator that performed threshold segmentation and noise filtering on flower cluster images, positioning the centroid of flowers as the spraying target, achieving an average recognition rate of 79% and a spraying success rate of 69.6%. Oppenheim et al. [13] provided an image analysis and detection algorithm for yellow tomatoes, suitable for greenhouse drone pollination and yield estimation tasks under unfavorable conditions such as uneven lighting, complex growth, and varying flower sizes.

Yahata et al. [14] proposed an image sensing method combining Simple Linear Iterative Clustering (SLIC) and hue information for detecting candidate regions, followed by Convolutional Neural Network (CNN) to assess flower receptivity, achieving coarse and fine flower detection. Dias et al. [15] developed a CNN + SVM method for apple flower recognition and detection, demonstrating high segmentation accuracy and generalizability to other flower types. The "BrambleBee" system uses fisheye and RGB-D cameras for initial flower recognition and precise positioning, ensuring detailed assessment of flower posture [6]. Kyoto University introduced an ultrasonic focused pollination device utilizing 3D cameras for strawberry flower color imaging and ultrasonic radiation to locate and vibrate flowers, achieving 93.9% recognition accuracy [16, 17]. Lim et al. [18] compared Faster R-CNN + NAS, Faster R-CNN + Inception V2, and SSD + Inception models for real-time agricultural pollination robots, achieving 91.9% detection accuracy on actual datasets. Shang et al. [19] utilized YOLOv5s deep learning to accurately detect apple flowers under various weather conditions and lighting scenarios. Hiraguri et al. [20] addressed the challenge of blurred images caused by drone and robot vibrations during pollination, developing a CNN-based image classification system enhanced with Gaussian filters for image smoothing.

However, current studies predominantly concentrate on flower recognition and detection, with limited focus on estimating flower posture. Estimating flower posture is crucial as it enables the end-effector to identify the optimal angle for pollination. Ohi et al. [6], in his subsequent work on BrambleBee, emphasized the need to enhance classifiers to more accurately estimate flower posture. This improvement is expected to enhance the effectiveness of the visual servo mechanism of the robot arm. Moreover, the capability to reliably distinguish individual flowers within clusters, even when they overlap, will further enhance pollination accuracy.

Table 1 provides the RGB camera, fish-eye camera, binocular camera, depth camera and other camera equipment required for flower recognition and positioning research, using image processing, deep learning and other methods for recognition, searching for centroid and other methods for positioning, as well as the corresponding accuracy or error rate. This information provides an at-a-glance look at the development so far, with detailed information available through the references cited; It can also be used as a tool to compare methods and peek into future research potential to further develop methods for identifying and locating different flowers. It can be found that deep learning is the future direction of flower detection.

2.3.2 Motion Planning of Pollination Robots

Autonomous navigation of pollinating robots is essential for enabling these robots to perform autonomous pollination and other operations without human intervention. Broadly, robot navigation encompasses positioning, map construction, and path planning. Positioning is the initial step, where the robot determines its location by sensing the surrounding environment using built-in sensors such as lidar, cameras, and ultrasound, and calculating its relative distance from nearby targets. In the Environmental Perception section, we discussed the application of various navigation modes in motion planning. Multi-sensor fusion, which combines data from different sensors, can significantly enhance the stability and accuracy of navigation systems, marking a key trend in future developments. Mapping, the second step, involves converting environmental data into a map. This process is crucial for achieving precise positioning and path planning, which are vital for safe and accurate navigation in complex and extreme environments. Environmental modeling bridges the gap between sensing and planning optimal paths. In the Environment Modeling section, we introduced Metric Maps, Topological Maps, mixed Maps, and Semantic Maps, and detailed the application of SLAM technology. Path planning techniques are categorized into global and local methods. Global path planning, or static planning, uses the robot's environment perception to follow a preset path to the target. Local path planning is suited for scenarios where the environment is partially or wholly unknown, enabling dynamic path planning or obstacle avoidance through real-time monitoring and response. Obstacle avoidance planning, which considers both space and time, is the most complex aspect of path planning. Practically, a combination of global and local methods is often used to achieve the global optimal path while ensuring real-time obstacle avoidance [27]. Path planning generally refers to finding a safe, collision-free, optimal path with the shortest distance and least time from the start to the endpoint, depending on the algorithm used. In the Route Planning section, we analyzed the strengths, weaknesses, and future trends of mainstream path planning algorithms. Figure 3 shows the system structure diagram of motion planning.

Environmental perception

Currently, widely employed navigation methods include satellite positioning navigation, laser positioning navigation, visual positioning navigation, inertial navigation, and multi-sensor fusion positioning navigation. GNSS positioning navigation demonstrates significant advantages and potential in open environments, making it suitable for expansive farmlands. Wei et al. [28] proposed a path search method for

Table 1 Analysis of various flower recognition and positioning technologies

Author(s)	Detection method(s)	Accuracy (%)	Location method(s)	Error	Camera(s)
Kim et al. [12]	Image processing	95.00	–	–	–
Hong and Choi [21]	Image processing	94.80	–	–	RGB camera
Yuan et al. [5]	Image processing	87.50(max)	The centroids of the target flower clusters	6.40 mm	RGB camera
Oppenheim et al. [13]	Image processing	74.00	–	–	RGB camera
Yahata et al. [14]	SLIC + CNN	90.00	–	–	RGB camera
Dias et al. [15]	CNN + SVM	92.70	–	–	RGB camera
Ohi et al. [6]	Naive Bayes' classifier Google's Inception-v3	78.63	Three-dimensional reconstruction	–	Fisheye camera RGB-D camera
Shimizu and Sato [17]	Image processing SVM	93.90	–	–	Stereo camera
Lim et al. [18]	Faster R-CNN NAS Faster R-CNN Inception V2 SSD Inception V2	96.80 90.40 78.50	–	–	Stereo camera
Li et al. [22]	Yolov4	96.60(flowers) 98.57(buds)	–	–	RGB camera
Hulens et al. [23]	SSD MobileNetV2	77.00	–	–	RGB camera
Shang et al. [19]	Yolov5s	97.20	–	–	RGB camera
Hiraguri et al. [24]	Image processing + CNN	87.30	–	–	RGB camera

(continued)

Table 1 (continued)

Author(s)	Detection method(s)	Accuracy (%)	Location method(s)	Error	Camera(s)
Li et al. [25]	Yolov4	93.00	The midpoint of the centerline connection between the flower outline and the pistil		Binocular camera
Mu et al. [26]	Mask R-CNN	78.10	Peak of super pixel density	5.40%	ZED-2 camera

Note SLIC simple linear iterative clustering, *CNN* convolutional neural network, *SVM* support vector machine', *SSD* single shot MultiBox detector. SSD MobileNetV2 is a lightweight object detection model that combines SSD (single shot multibox detector) and MobileNetV2

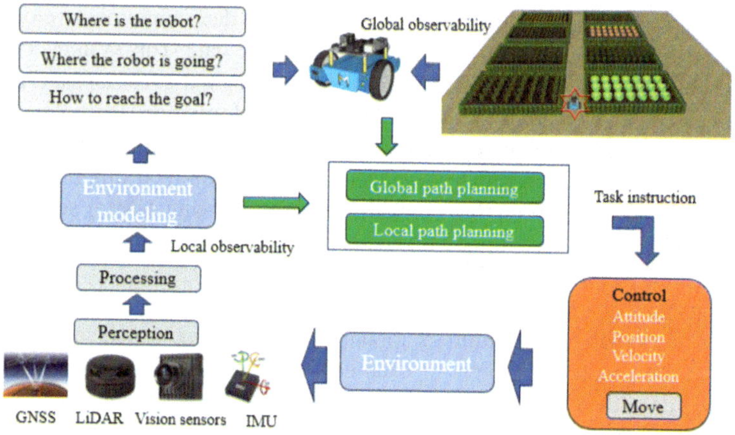

Fig. 3 Robotic pollination related publications: **a** number of publications for individual countries, **b** CiteSpace map of different publication countries, **c** number of publications per year and **d** related areas to robotic pollination studies

agricultural machinery based on GNSS, employing a pure tracking model with pre-sight point search. Han et al. [29] developed a path generation and tracking system for an automatic driving sprayer using single-frequency GNSS PTK, achieving positioning accuracy up to 0.01 m. Optical ranging and LiDAR ranging have also become reliable technologies for 3D environment awareness, extensively utilized in autonomous driving for dynamic and static scene analysis, mapping, and situational awareness. Their functionality extends beyond obstacle detection [30]. Jones et al. [31] designed the Macaque Peach Garden autonomous navigation platform based on multi-line LiDAR. Wu and Li [32] employed a four-wheel differential robot as a test platform, utilizing single-line LiDAR to obtain environmental data. They processed laser data using distance threshold methods, adaptive density clustering algorithms, and improved least square methods to derive tree-line centerlines as navigation paths for the robot. Experimental results demonstrated stable navigation between tree lines at speeds of 0.2, 0.4, and 0.6 m/s. Within certain offsets and deflection angles, the lateral deviation between the robot's navigation track and the tree-line centerline was minimal, with an average deviation of 3.69 cm and a maximum deviation of 11 cm. Furthermore, the robot executed accurate turns to transition between lines and continued autonomous line following during tests.

As image technology advances, vision sensors are increasingly integrated into agricultural robots for their capability to provide abundant information at low cost [33]. Radcliffe et al. [34] utilized vision sensors to extract tree crown and sky features, segmenting object centroids as process variables to guide unmanned ground vehicles along tree line centers. Their experiment demonstrated a root-mean-square error accuracy of 2.13 cm. Hai et al. [35] proposed a new full-convolutional network based on Retinex theory to enhance the quality of low-light images captured by drones, thereby improving visual quality and image matching performance effectively [35]. Inertial navigation systems autonomously determine robot velocity and position by measuring acceleration and integrating the data. This system comprises inertial measurement devices including accelerometers and gyroscopes, crucial for obtaining aircraft acceleration and angular velocity information. Inertial navigation systems exhibit robust autonomy, are resilient to external disruptions, and hold indispensable roles in various applications. Cai et al. [36] classified different gaits and paired them with corresponding Inertial Measurement Unit (IMU) sensors, integrating inertial navigation with Zero Velocity Update (ZUPT) for real-time estimation and correction.

A single satellite navigation signal is susceptible to obstruction and interference from obstacles such as tall buildings, mountains, and forests, which can degrade the quality of the positioning signal or even cause loss of location. Adverse weather conditions can affect the laser's effective range and introduce cumulative errors that degrade its performance. Relying solely on visual navigation in outdoor environments is vulnerable to lighting variations, while inertial navigation can accumulate significant long-term positioning errors. To address these challenges, in recent years, many researchers have adopted a multi-sensor fusion approach to enhance navigation system stability in complex environments. By integrating data from different sensors, algorithms can compensate for individual sensor limitations, thereby enhancing overall system stability and accuracy [22, 37].

Indoor environments and orchards pose challenges for GPS positioning due to complex conditions, signal interference, and multipath effects, which can degrade accuracy and adaptability. The proliferation of wireless infrastructure and mobile devices has spurred advancements in indoor localization, with received signal strength (RSS)-based localization emerging as an effective solution [38]. Common wireless signals utilized include radio frequency identification (RFID), Wi-Fi, Bluetooth, and ultra-wideband (UWB), each supporting location technologies. The accuracy of these systems relies heavily on sophisticated location algorithms such as direct positioning, geometric calculations, and fingerprint positioning. Measurement techniques pivotal to these algorithms include time of arrival (TOA), angle of arrival (AOA), time difference of arrival (TDOA), and RSS [39–42]. Zhou et al. [43] introduced the Q-WKNN algorithm, a weighted K-nearest neighbor (WKNN) approach based on RSS transformation for indoor positioning. Experimental results demonstrated that Q-WKNN outperformed WKNN, modified WKNN (M-WKNN), Gauss kernel (GK), and least square support vector machine (LS-SVM) algorithms in terms of accuracy and real-time performance. Moreover, employing interference positioning based on wavelength analysis of received waveforms can achieve high positioning accuracy within a few centimeters of error, representing a robust method [20]. Unmanned aerial vehicles (UAVs) leverage interference positioning for terrain following, target tracking, and automatic landing, while ground robots utilize this technology for map construction, path planning, and target positioning, ensuring precise, high-resolution, and interference-resistant positioning and navigation capabilities.

Environment modeling

Environmental modeling abstracts real physical space into an abstract space that algorithms can process, establishing an environment model conducive to computerized path planning. Mapping is a specific operation within environmental modeling, where the resultant models typically manifest as maps. Each element in these maps corresponds to an object or terrain feature in the actual environment. Through maps, path planning algorithms gain enhanced understanding and operational capability of the environment, facilitating more effective planning of optimal paths. Various methods exist for map construction, including grid maps, octrees, and voxel hashing. Common types of maps include metric maps, topological maps, hybrid maps, and semantic maps, each serving distinct purposes in facilitating efficient path planning and navigation.

A metric map describes the environment using mathematical measurements, typically represented using 2D or 3D grid maps, emphasizing precise representation of object relationships, often classified as dense or sparse (Fig. 4a). Both 2D grid maps and 3D point cloud maps fall under the category of metric maps. In a topological map, each node corresponds to a distinct location in the real environment. A hybrid map integrates components and characteristics of both metric and topological maps, incorporating topological information into metric maps or refining points in topological maps using metric map data (Fig. 4b and c) [44]. Semantic maps provide qualitative descriptions of the robot's surrounding environment, enhancing navigation capabilities, task planning abilities, and bridging gaps in human–robot interaction (Fig. 4d). Wu et al. [45] addressed the instability of traditional SLAM systems in dynamic environments by utilizing the Darknet19-YOLOv3 object detection network to accelerate and generate necessary semantic information. They differentiated dynamic features within detection regions using depth consistency of random samples to mitigate the impact of dynamic targets [46]. Sengupta et al. ([47], 2015) generated semantic labels for street images by directly segmenting frames. They aggregated the output results from multiple consecutive frames to create a large-scale semantic map.

Lu et al. (2017) categorized map construction into indirect methods, direct methods, mixed methods, and multi-sensor fusion. Indirect methods, which extract features from images and use them as inputs for motion estimation and positioning processes, are prone to issues in untextured environments. Direct methods, which

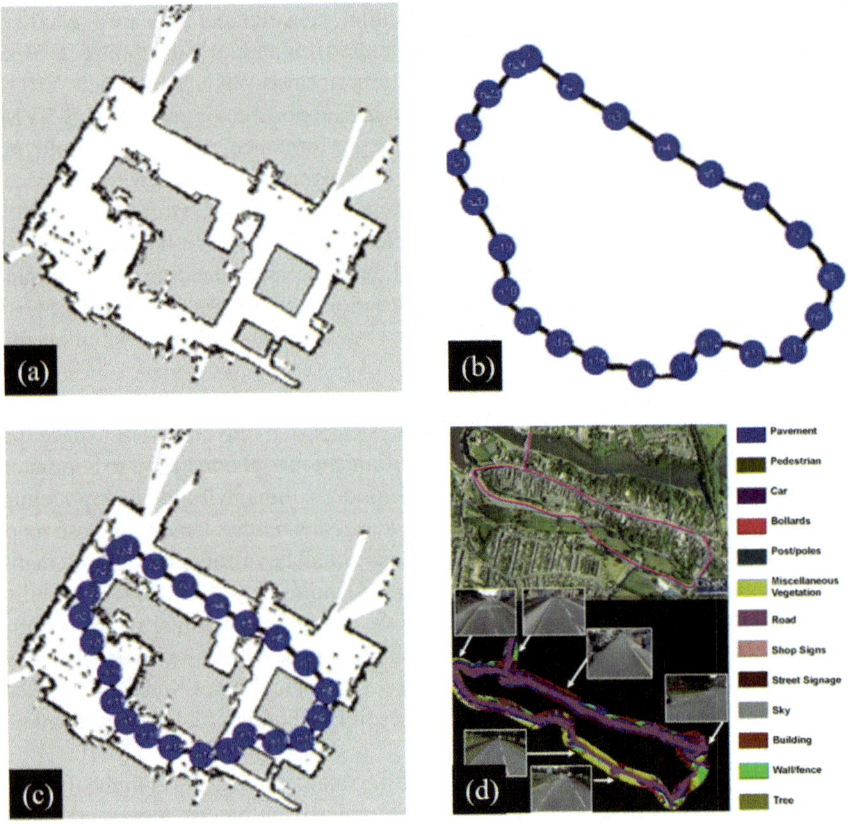

Fig. 4 A 4 methods of building a map: **a** metric map; **b** topological map; **c** hybrid map and **d** semantic map

utilize all available information in the image, are more reliable in environments with few feature points and have been a focus of research in composition systems. The hybrid method combines these two approaches to continuously adjust the camera position, resulting in faster and more accurate performance [48]. Lynen et al. [49] developed a universal multi-sensor fusion Extended Kalman Filter (MSF-EKF) capable of processing various types of delayed measurement signals from multiple sensors. Simultaneous Localization and Mapping (SLAM) technology represents a form of environmental modeling where, without prior environmental information, the subject equipped with specific sensors establishes an environmental model while simultaneously estimating its own movement. Currently, the mainstream SLAM technologies are primarily laser SLAM and visual SLAM. SLAM can provide real-time and high-precision positioning and navigation information for robots. Magree and Johnson [50] developed an integrated navigation system that combines visual SLAM and laser SLAM with an inertial navigation system based on the EKF. Monocular Visual SLAM is responsible for data correlation and estimation of UAV status in the system, while laser SLAM handles scanning and map matching within the Monte Carlo framework. SLAM is increasingly being applied to research in orchard and greenhouse navigation [22, 37].

"BrambleBee" adopts a real-time 3D-SLAM algorithm, using the factorization of the posterior distribution to assume that the system follows a Gaussian noise model. This simplifies the state estimation problem to a nonlinear least squares optimization problem. The GICP unified model matches the information collected by the 3D LiDAR with the 3D map in the room to improve the reliability of loop closure [6]. Traditional visual SLAM systems face performance issues in dynamic environments. To address this, Zhang et al. [51] proposed a semantic SLAM system specifically designed for the use of refractive and reflective panoramic cameras in dynamic environments. This system employs a real-time instance segmentation network to detect potential moving targets, verifies them based on pole constraints of the sphere, and then uses only static feature points to estimate camera attitude. Experimental results demonstrate that the system's accuracy is 96.3% higher than traditional algorithms in highly dynamic environments.

For pollinating robots, the surrounding environment is complex, necessitating continuous improvement in the accuracy and stability of the sensing system. Additionally, flower detection can be integrated with 3D-SLAM technology to create 3D semantic maps of flower cluster locations. The detected stages of flower development can be stored together with flower locations, enabling the planning of autonomous pollination tasks over multiple days [6].

Route planning

Path planning is a typical NP-hard problem, involving multiple targets whose solution requires polynomial verification. Path planning aims to assist robots in completing tasks with higher reliability, faster speed, and lower energy consumption [52, 53]. Path planning algorithms can be divided into three main categories: classical algorithms, bionic algorithms, and artificial intelligence algorithms. Classical algorithms do not require expensive environmental monitoring equipment and have a small memory footprint, making them suitable for global path planning of low-cost mobile robots. However, they suffer from issues such as a lack of intelligence, susceptibility to local optimal solutions, and low computational intensity, which limits their real-time path planning capabilities. Examples of classical algorithms include the Cell Decomposition Approach (CD), Sampling-Based Methods (SBM), and Graph Search Algorithms (GSA), which exhibit these performance characteristics. The Artificial Potential Field (APF) algorithm combines the advantages of both global and local path planning. Artificial intelligence algorithms, due to their strong learning ability, can effectively handle environmental uncertainties, providing significant advantages [54, 55]. In the realm of bionic algorithms, Liu et al. [55] provided a detailed literature review covering global optimization, local optimization, dimensionality, smoothness, and other aspects, clearly identifying directions for algorithm optimization and integration. Yang et al. [56] explored the challenges of Single-Agent Path Planning (SAPF) and Multi-Agent Path Planning (MAPF). SAPF focuses on determining the optimal path for a single robot to navigate from its initial position to a specified target. MAPF, however, requires coordinating multiple robots in the same environment to ensure they complete their tasks without conflict. These studies offer valuable insights and inspiration for advancing the field of path planning.

Most algorithms struggle to meet real-time requirements in complex environments. For instance, Cell Decomposition (CD) primarily focuses on optimizing the algorithm itself, while Sampling-Based Methods (SBM), Cuckoo Search Algorithm (CSA), and Artificial Potential Field (APF) similarly emphasize algorithmic improvements. Particle Swarm Optimization (PSO) and Genetic Algorithm (GA) are often combined with other algorithms to enhance performance. However, it is notable that current research tends to prioritize algorithm improvements over fusion and pays less attention to path smoothing techniques. As robot technology continues to advance, autonomous path planning will encounter increasingly complex and variable environments. This necessitates algorithms that can rapidly adapt to complex environmental changes, making it clear that a single algorithm cannot address all challenges. Consequently, future practical applications will likely involve the combination of global and local path planning, the integration of traditional and artificial intelligence methods, multi-sensor fusion, and multi-robot coordination planning. Wang et al. [57] proposed a neural network-driven prediction method (NEED) that predicts regions where viable paths are likely to exist by learning the environmental structure. It features an encoder-decoder architecture composed of three modules: a convolutional neural network backbone, a spatial pool module, and a decoder module. Three-dimensional path planning is more complex and less studied than two-dimensional path planning, with most research limited to simulations. Yi et al. [58] combined the features of Bayesian Neural Networks (BNN) with self-organizing feature mapping to handle task assignments for a large number of robots in a 3D dynamic environment. For 3D path planning, Song et al. [59] proposed an improved Cuckoo Search Algorithm (CSA) based on compact parallel technology.

2.3.3 Current Research on Pollination End-Effectors

The end-effectors used for pollination are less diverse than those used for picking. Pollination end-effectors can be broadly categorized into three types: spray pollination, vibration pollination, and contact point pollination. Each type has its own characteristics and application scenarios to meet various pollination needs, serving as extensions of traditional manual pollination methods. Spray pollination has been extensively used in pollination research due to its high efficiency.

Raz Sprayers, an Israeli company, manufactures sprayers for vineyards, greenhouses, date palms, and other applications. For instance, Fig. 5a-I depicts a hydraulic lifting sprayer for spraying and dusting date palms, operated from the driver's seat [60]. Gan-Mor et al. [9] designed a stable mobile robot arm for pollinating tall trees. Stability tests on the robot arm, conducted on a test track, indicated that when the frame tilt reached 10 degrees, the main arm's tilt remained within 2.6 degrees, and the prediction deviation was within an acceptable range, enabling precise pollen spraying in jujube orchards. Israel has also developed a visually guided date palm sprayer with a spray head comprising a video camera and a spray hose on a telescopic mast, operated by a winch on a towing platform. Figure 5a-II shows the end picture of the sprayer in action [61, 62]. Williams et al. [63] introduced a new type of kiwi pollinator robot that performs pollination using an air-assisted spray tube, shown in Fig. 5a-III. The spray tube has two cameras at the front 300 mm and white light bars (6000 K) on both sides to mitigate ambient lighting effects. The tube, comprising 20 nozzles, transports pollen solution to flowers located 400 mm above the nozzle to complete pollination. However, the pollination success rate of this pollinator was only 79.5%, and the quality of the resulting fruit did not meet commercial standards [63–66]. Figure 5a-IV illustrates the greenhouse tomato hormone pollinator robot mentioned in Sect. 2.3. The specially developed end effector includes a reagent spray nozzle, a spray control solenoid valve, and a manual valve for water control. After detecting flower clusters, the pollinator robot moves to the target flower and sprays the configured pollination solution through the solenoid valve onto the flower [5].

This type of spray pollination is suitable for various plant types, especially for large-scale crops. However, caution is necessary regarding the force applied during spraying, as excessive jet force can cause mechanical damage to the flowers. Additionally, spray pollination can result in problems such as low hit rates, wasted pollen, and environmental pollution due to extensive spraying. In recent years, electrostatic micro-spraying has gained popularity in agriculture. By utilizing the principle of high-voltage static electricity, a high-voltage electrostatic field is created between the

Fig. 5 End effector for mechanical pollination: **a** Spray pollination end effector; **b** Vibration type end effector; Contact type end effector

sprinkler head and the crops. When the pollen solution passes through the sprinkler head, the electrostatic field charges the droplets, enabling uniform spraying. Bechar et al. [67, 68] designed a machine that charged pollen and transported it to flowers, improving electrostatic deposition technology. Field tests in California demonstrated that electrostatic pollination increased pistachio yields by 11.3%, with pollen density on the pistil after electrostatic pollination being 30 times higher than on the envelope. Bechar et al. [69] developed an electrostatic ring (ESR) to enhance pollen penetration through the tree layer and reduce the impact of the distance between the electrostatic pollinator and the tree on pollen penetration and deposition. These advancements demonstrate the potential of developing atomized spraying and electrostatic pollination technology.

Vibrators are generally suitable for self-pollinating plants as they do not directly touch the flowers, thereby reducing the risk of mechanical damage. Shimizu and Sato developed an ultrasonic pollination device that combines a 3D camera with an ultrasonic focusing device. This device directs ultrasonic waves onto the flowers, causing them to vibrate and complete pollination [16, 17]. As illustrated in Fig. 5b-I, the pollination device remains stationary during pollination, while the cultivation bed moves. This device was tested on a strawberry farm in Ehime Prefecture, Japan, and the results showed that the marketable rate of "Akai Shizuku" strawberries under 40 Hz ultrasound could reach 91.7%. However, the device has stringent requirements for specific pollination varieties and greenhouse cultivation, necessitating cultivation rules and movable cultivation beds, which pose challenges for commercialization.

The end effector of the contact point pollinator can precisely align with the flower to ensure direct pollen delivery, which enhances the pollination success rate. Figure 5c-I depicts a pollination platform designed for vanilla plants, featuring a docking post and two robotic cranes on either side. These cranes control the platform's movement, supported by a single structure that can accommodate two to six such systems working in parallel [70]. A significant challenge in contact pollination is obtaining pollen. Yu et al. addressed this by integrating the TFDC-Net (Tomato Flower Detection and Classification Network) model into a pollinator robot, as shown in Fig. 5c-II. This model incorporates target detection, flower classification, and pose recognition algorithms, utilizing a depth camera on the robot arm to locate and identify flowers in the field of vision. It determines the flowering stage and posture, calculating the position of fully bloomed flowers with consistent posture. The end-effector then collects pollen and performs contact pollination with the polygonium [71]. Before and after the modification of "BrambleBee," the end effector comprised a depth camera, an endoscopic camera, a server, three linear actuators, and a flexible member with pollination swabs inserted. As shown in Fig. 5c-III, the flexible component forms a concave shape relative to the flower when the linear actuator is extended, allowing it to friction with the anther. When the actuator is retracted, the flexible component protrudes, enabling pollination [6]. The improved end-effector added a layer of cotton wool to the front section of the flexible component to increase the contact area between flowers and pollen (Fig. 5c-IV) [7]. However, contact pollination operations are complex, requiring a certain level of technical expertise, making it challenging to improve pollination efficiency. Consequently, it has not been widely adopted in production.

Currently, most pollination end-effectors are still in the experimental stage. There are two main challenges: improving the generalization and accuracy of spray and vibration pollination, and enhancing the efficiency of contact point pollination.

2.4 *Drone Pollination*

There are two fundamentally different approaches to robotic pollination. The first approach involves using a platform equipped with a robotic hand and a sprayer, as mentioned above. The second approach involves employing drones or robotic bees [72]. With the miniaturization of sensors, computing modules, and driving mechanisms, the design of new micro-aviation (MAV) and nano-aviation (NAV) aircraft has become feasible, prompting many scholars to combine bionics and microrobotics [73–75]. Berman et al. from Harvard University, proposed swarm control and coverage strategies for flying robots, applying these methods to pollination (Fig. 6a) [76–78]. Abutalipov et al. [79] also suggested the use of nanocopters (UAVs) as robotic pollinators. In 2018, Walmart filed a patent for autonomous robot bees, aiming to enable these tiny drones to pollinate crops autonomously, similar to natural bees [80]. The Light Robotics Group at the University of Tampere developed a polymer assembly robot that flies with the wind and is controlled by light (Fig. 6b) [81]. Yang et al. [82] a member of the team, envisions a future where millions of pollen-carrying artificial dandelion seeds are dispersed by natural winds and directed by light to specific areas of trees awaiting pollination. Priscilla et al. [83] utilized clustering algorithms to find the center of mass and determine flight direction, allowing robobee's wings and body to be regularly updated for intelligent interaction based on artificial intelligence, addressing issues such as insufficient activity and unidirectional flight. However, some scholars remain skeptical. Despite rigorous modeling of aircraft pollination methods, there is currently no actual design or physical system to verify their effectiveness. The Harvard team's robobee, a bionic robot under development, still requires fixed power and control, making it far from becoming a viable pollinator. Additionally, there are ecological and moral concerns. Potts et al. [84] argue that the use of biomimetic robots for pollination could undermine biodiversity, potentially leading to ecosystem destruction and food insecurity.

In addition to bionic robots, rotorcraft drones are widely used in precision agriculture due to their size and stability [85]. These drones are employed in tasks such as aerial crop monitoring and spray-only missions, and significant advances have been made in pollination research. Utilizing advanced flight control and sensor technology, rotorcraft can accurately locate and pollinate flowers. For instance, the U.S. startup Dropcopter has developed an autonomous drone capable of achieving a pollination rate of 25–60% [86]. Miyako developed a small UAV for artificial pollination (Fig. 6c). This drone, equipped with sticky gel and horsehair on its abdomen, collects

Fig. 6 Micro robots or UAV (Unmanned Aerial Vehicle) for pollination

and releases pollen, achieving a 53% success rate in pollen collection and a 37% success rate in pollination. However, it operates manually rather than autonomously and can damage flowers during the process [87]. To address this issue, Miyako and Yang [88] proposed using air bubbles for pollen transport. In laboratory tests, they bombarded pear flowers with pollen-filled bubbles, which burst upon contact, releasing pollen onto the pistil and achieving pollination. This method resulted in a pollination rate of about 95%, with the fruit quality comparable to that of artificial pollination. To demonstrate the scalability of their bubble pollination method, they equipped a drone with a bubble machine capable of generating 5000 bubbles per minute. When the drone moved over flowers at 2 m/s at an altitude of 2 m, the overall success rate was 90%. However, in actual operation, many bubbles failed to reach their target, leading to wasted pollen. Additionally, due to the light mass of the bubbles, controlling their trajectory is challenging even with enhanced recognition technology. Furthermore, the bubble pollination method raises concerns about chemical interference with pollination and potential soil contamination (Fig. 6d).

The UAV pollination system must not only meet the requirements outlined in Sect. 2.2 but also possess the following characteristics: (a) lightweight; (b) low power consumption; (c) ease of deployment and integration; and (d) a high frame rate to meet real-time requirements. Hulens et al. [23] developed a navigation system for pollinating drones that addresses these needs and conducted experiments to validate its performance. The results demonstrated that the drones could autonomously detect flowers, estimate angles, and fly to the flowers for pollination, achieving an 87% success rate (Fig. 6e) [89]. However, the system's flower detection efficiency is not sufficiently high, and its accuracy may decrease when used on smaller flowers, as it was tested on larger sunflower blooms. Hiraguri et al. [24] integrated an artificial intelligence flower detection system, autonomous flight control for flower searching, and a pollination control device on a small UAV, enabling it to pollinate greenhouse-cultivated tomatoes. They used multiple infrared cameras in the greenhouse with motion capture technology to locate the drone within the environment by detecting reflected markers attached to it and calculating their 3D coordinates. Flower image data taken by the drone were used to identify flowers suitable for pollination, classified by an AI classifier, and shared via Wi-Fi. The flight control node then issued flight commands based on the classification results. Experiments showed that drone pollination by vibration had a significantly higher fruit setting rate than hormone pollination and was comparable to bumblebee pollination. However, motion capture technology is rarely used for positioning due to its lack of accuracy. Mazinani et al. [90, 91] designed a pollen tank with holes for releasing pollen (Fig. 6f). They also used Ansys Fluent software to simulate the Computational Fluid Dynamics (CFD) of the airflow path of a quad-rotor UAV under three conditions: hovering, moving forward, and with wind. The results indicated that the pollen path changes with wind disturbance, complicating scheduled pollination. In UAV flower pollination research, flower perception and path planning remain crucial, and UAV flight control methods require further study. Rice et al. [92] introduced an autonomous UAV-based pollination system focusing on flower perception, path planning, and flight control. They compared two path planning methods: mixed integer programming (MIP) and a genetic algorithm. The genetic algorithm proved to save significant time and computational resources. Rice also proposed a convex optimized flight control method and conducted numerical verification, finding that with 10 nodes, the solution generation speed was 15 times faster than that of GPOPS (General Pseudo-Spectral Optimal Control Software). However, the advantage diminishes with a larger number of nodes.

With the advancement of liquid pollination technology, the use of knapsack sprayers for pollination has matured, leading researchers to explore UAV spray pollination technology. Wang set three different spray amounts for multi-rotor UAVs to study the effects of spray parameters on droplet size, deposition amount, and fruit setting rate. The experimental results showed that droplet size and coverage on the upper layer of pear trees were higher than on the lower layer, which contrasted with knapsack sprayers. The droplet size and coverage of UAV spray pollination were positively correlated with the spray amount. Higher spray volumes resulted in better pollination effects, though still inferior to manual point pollination and emerging knapsack sprayer liquid pollination [93]. Wang's similar study showed that under conditions of wind speeds less than 2.0 m/s, UAV flight height 1 m above the tree top, nozzle atomization particle size 0.120 mm, and flight path above the tree top, dispersal of stratified Korla Xiangpear with 2.0 L/667 m^2 achieved the best pollination effect (Fig. 6g) [57, 94]. Jia et al. [95] proposed using thermal infrared images to accurately identify pollination areas through floral thermogenesis. They achieved virtual simulation of precise pollination using a 4-unit multi-rotor UAV cluster and performed timeliness analysis of the algorithm in a MATLAB environment. Currently, there are two main methods of rotary UAV pollination: liquid pollination and wind pollination. Wind pollination is mainly used for self-pollinating crops, such as rice. This method employs UAV rotor airflow to transfer pollen from male to female plants. Shi et al. installed deflectors on both sides of a drone to change the direction of the airflow downward, allowing the airflow to directly affect the tomato flowers and solve the issue of UAVs being unable to fly over tomatoes due to sling attachments during the flowering stage (Fig. 6h) [96]. Compared to manual pollination, UAV spray pollination offers higher efficiency and more uniform fruit setting, but its fruit setting rate remains inconsistent and is significantly affected by environmental factors. Thus, further research is necessary to develop precise pollination methods for fruit trees.

3 Field Crop Pollination

As a major global food crop, rice seed production is a crucial aspect of its overall cultivation process. Adequate and uniform pollination can significantly enhance the seed setting rate, which is essential for improving both yield and quality of seed production. Auxiliary pollination is one of the key factors to ensure successful seed production. The assisted pollination of hybrid rice demands high technical precision and timely execution, with pollination effects being highly susceptible to environmental conditions. Unlike previously described pollination methods, field pollination methods for rice are distinct and thus require a specific summary. Field pollination methods can be divided into two categories: artificial pollination and mechanical pollination. Manual pollination is a well-established method with reliable results; however, it is labor-intensive and restricts parent plants to small, alternate rows, hindering mechanized operations. In contrast, research on mechanical assisted pollination is advancing rapidly in China, leading to the deployment of various pollination machines, including UAVs. Despite these advancements, there is no specialized pollinator designed for large row ratios, and fundamental research on pollination in large row ratio planting modes is scarce. This lack of basic research limits the theoretical support needed for optimizing mechanical pollination in these planting configurations [97].

3.1 *Artificial Pollination*

In the practice of seed production, three primary methods of artificial pollination have evolved based on regional and customary differences: the rope pollination method, the single rod pollination method, and the double rod pollination method. These labor-intensive techniques were established in the mid to late 1970s.

Rope Drive Method: This method is the most widely used and efficient for hybrid rice seed production. During the pollination process, two staff members pull the ends of a rope, moving in parallel with the rice row and pulling it back and forth. This action scatters the pollen from the male plants into the female rows, thereby completing artificial pollination (Fig. 7a).

Single Rod Pollination: This method includes two variants: vibration and push.

Fig. 7 Different auxiliary pollination methods in the field: **a** pollination by one pole vibration, **b** POLLINATION by two poles, **c** pollination by ropes, **d** pneumatic hybrid rice pollinator, **e** self-propelled pollinator with high ground clearance, **f** pipeline airflow pollinator, **g** artificial backpack pollinator, **h** drone pollination

(i) **Vibration Method:** One person holds a rod and shakes the bottom of the male ear layer to scatter pollen into the female plants. This method generates a high vibration force, dispersing pollen far but unevenly.

(ii) **Push Method:** The rod is held in the middle and upper part of the male plants and pushed forward to scatter pollen evenly over the female plants, avoiding damage to the female plants. The rod is then pushed again on the return pass (Fig. 7b).

Double Rod Pollination: In this method, the operator places two bamboo rods in the middle and upper parts of two rows of male plants. The rods are then rapidly vibrated 2–3 times, ensuring that the male pollen is fully dispersed and spread to the female rows on both sides. The double-rod push pollination method allows for more thorough pollen dispersion, longer propagation distances, and more uniform pollen distribution (Fig. 7c).

Table 2 illustrates the parent-to-parent planting ratios, parent row widths, and pollination efficiencies for three artificial pollination methods. The rope powder drive method, requiring 3–4 operations daily, is characterized by high speed and efficiency. Two people can pollinate an area of 0.7–1.3 hectares per day. However, this method results in uneven pollen dispersal and a low utilization rate, making it better suited for male single-row and small double-row sterile lines or hybrid seed fields. The single pole pollination method is slower, allowing one person to cover only 0.2–0.3 hectares per day, which is time-consuming and labor-intensive, leading to its decreased usage in current production practices. The double rod drive method, appropriate for breeding male sterile lines or hybrid seed fields with large double-row or small double-row male cultivation, enables one person to pollinate only 0.07–0.13 hectares per day. This method is also labor-intensive and struggles with timely pollination during the peak of male flowering. The narrow planting ratio complicates mechanized planting, pollination, and harvesting of rice. Therefore, increasing the

Table 2 Comparison of artificial assisted rice pollination methods

Methods	Interplanting ratio of parental lines	Paternal row spacing (cm)	Mater row spacing (cm)	Pollination efficiency
String pollination method	1 ~ 2:10 ~ 12	60 (single-row) 70 (double-row)	160 ~ 185	0.7 ~ 1.3 hm^2/2person/day
Single-rod pollination method	1 ~ 2:10 ~ 12	60 (single-row) 70 (double-row)	160 ~ 185	0.2 ~ 0.3 hm^2/person/day
Double-rod pollination method	2:14 ~ 16	80	160 ~ 185	0.07 ~ 0.13 hm^2/person/day

parent-row ratio is crucial for improving seed production yield and achieving full mechanization.

3.2 *Mechanical Assisted Pollination*

Since the 1980s, China has been exploring mechanical equipment and technologies for hybrid rice seed production, including mechanical pollen collection, concentrated pollen spraying, auxiliary fan pollination, and knapsack pollination machines. The method of concentrated pollination allows for the planting of parent plants in centralized fragments, which can somewhat mitigate the issue of low parent planting ratios. During the father's blooming period, pollen is mechanically collected and stored at low temperatures, then diluted and sprayed during the mother's blooming period. This method addresses the problem of asynchronous flowering times between parent plants. However, experiments have shown that low-temperature pollen storage often leads to a loss of vitality and low germination rates, making it challenging to apply in practice [98]. To simplify the pollination process, pollinators prefer real-time pollination, which involves directly transporting pollen to the mother plant using external forces. This approach can be classified into three types based on different principles: collision type, pneumatic type, and UAV-assisted pollination.

The position of airflow has a significant influence on pollen quantity and distribution during pneumatic pollination. Analyses of plant stress and pollen movement under airflow showed that optimal pollen diffusion occurs when the airflow center targets the middle of the ear head [99, 100]. Wu et al. [101] invented a collision-pneumatic hybrid rice seed pollinator that combines collision and air blowing for pollination. Yao [97] explored the effects of pneumatic pollinator tube parameters, such as airflow outlet velocity, airflow action position, and forward velocity, on the head displacement, velocity, and acceleration of male hybrid rice. They developed a pneumatic hybrid rice seed pollinator suitable for large row ratio planting, determining the optimal combination of airflow parameters for each paternal row

(Fig. 7d). In the United States, after the introduction of hybrid rice from China in the 1980s, mechanized seed production technology was explored. By the mid-1990s, Circle S Seeds-Grain company used small manned helicopters for assisted pollination. The helicopter rotors' wind speed propagated male pollen to the mother plants, achieving a matching parent ratio of (8–10): (30–40), with male chamber widths of 2–2.4 m and female chambers of 9–12 m, enabling mechanized planting and harvesting. In China, the application and promotion of plant protection UAVs began in 2012, initially focusing on plant protection rather than hybrid rice seed production's assisted pollination mechanization. Su compared the pollination effects and efficiency of a fully autonomous four-rotor plant protection UAV, a high-gap self-propelled pollinator (Fig. 7e), and a manual knapsack pollinator (Fig. 7g). The piped-type airflow pollinator installed on the self-propelled pollinator assisted pollination. Results indicated that both the upland gap self-carrying pollinator and the manual piggy-back pollinator could achieve or exceed the plant protection UAV's pollination level (Fig. 7e–g, Su 2019). Weng et al. [102] compared UAV-assisted pollination and rope-driven pollination regarding pollen density, seed setting rate, work efficiency, and cost. Both methods had similar pollination effects, but drone-assisted pollination saved 50% of manpower and was over ten times more efficient than the rope method (Fig. 7h). In recent years, drone-assisted pollination has gradually entered the application stage, with farmers in Hunan, Sichuan, and other regions using drones for rice pollination. Experiments proved that even in natural wind conditions of level 4, unmanned helicopter-assisted pollination could match the effect of manual pollination. However, issues like poor autonomy and pollen drift remain, necessitating further research to improve effectiveness. The wind field generated by rotary-wing UAVs significantly affects rice pollen distribution, directly impacting hybrid rice breeding [103]. Studies by Zhejiang University, South China Agricultural University, and Jiangsu University on UAV flight height and speed showed effects on droplet deposition density, coverage rate, wind field distribution, and deposition uniformity, verifying crucial data such as flight height, speed, spray width, and pollen density [104–108].

4 Summary and Conclusions

In this study, the main aspects of assisted pollination were reviewed, focusing on fruit setting rate and pollination efficiency. Pollination is a crucial link between production and management, and it should be an integral concept and application within "smart agriculture." An efficient crop pollination platform needs to derive information from the vast data generated by smart agriculture and provide important dimensional information for it. However, current academic research and support for mechanized pollination remain insufficient. The goal of mechanized pollination is to achieve efficiency and precision [46]. Despite significant progress in mechanized pollination over the past 20 years, many challenges persist. These challenges include system adaptability, technical knowledge, national infrastructure, fabrication scale, and the overall economics of future research and method development. While artificial pollination has high costs and can harm workers, it offers the highest fruit setting rate and serves as a fallback when mechanized pollination fails. The lack of real-time hardware and software performance limits the efficiency and accuracy of system operation. Additionally, current pollinating robots are confined to greenhouses with flat and regular ground due to technological constraints. While rotorcraft drones have successfully applied pollen solution spraying in practical scenarios, significant breakthroughs in bionic robots and small rotorcraft drones for pollination remain elusive. Drones with axle diameters under 500 mm or bee-sized drones struggle to hover stably in strong winds and face challenges in power supply and pollen carriage. Building power supply piles or using mobile chassis cars for real-time power supply to drones is a potential solution, but cost considerations are significant. Furthermore, research on simultaneous pollination of cross-pollinated plants is scarce. High-frequency vibration presents another challenge for stable and accurate image recognition, requiring cameras to automatically adjust exposure time and white balance and take clear photos under such conditions. After describing several mainstream assisted pollination methods, this paper reviews the pollination status of field crops in unique application scenarios. Drone pollination is expected to bring more efficient development. As an essential component of smart agriculture, advancements in mechanized pollination will revolutionize agricultural production, enhancing efficiency and productivity.

References

1. Pollen Plus (n.d.) Quadduster. https://www.pollenplus.co.nz/quadduster
2. Chen Y, Li Y (2019) Intelligent autonomous pollination for future farming—a micro air vehicle conceptual framework with artificial intelligence and human-in-the-loop. IEEE Access 7:119706–119717. https://doi.org/10.1109/ACCESS.2019.2937171
3. Ma J (2010) Effects of different concentrations of 2,4-D dipped in flowers on the fruit setting rate of tomatoes in early spring sunlight greenhouses. Northern Hortic 22:58–59
4. Sun Y, Xu G, Gao W et al (2017) The effects of different pollination methods on the yield, quality, and benefits of tomato in solar greenhouses. Chin Vegetable J 06:38–41. https://doi.org/10.19928/j.cnki.1000-6346.2017.06.008
5. Yuan T, Zhang S, Sheng X et al (2016) An autonomous pollination robot for hormone treatment of tomato flower in greenhouse. In: 2016 3rd international conference on systems and informatics (ICSAI), Shanghai, China, pp 108–113. https://doi.org/10.1109/ICSAI.2016.7810939
6. Ohi N, Lassak K, Watson R et al (2018) Design of an autonomous precision pollination robot. In: 2018 IEEE/RSJ international conference on intelligent robots and systems (IROS), Madrid, Spain, pp 7711–7718. https://doi.org/10.1109/IROS.2018.8594444
7. Strader J, Nguyen J, Tatsch C et al (2019) Flower interaction subsystem for a precision pollination robot. In: 2019 IEEE/ RSJ international conference on intelligent robots and systems (IROS), Macau, China, 2019, pp 5534–5541. https://doi.org/10.1109/IROS40897.2019.8967752
8. Perkins R, Burkner PF (1974) Mechanical pollination of date palms. Calif Agric 28:6–7
9. Gan-Mor S, Grinshpon Y, Glik Y et al (2008) Stabilization of a mobile robotic arm for precise spraying and pollinating in tall trees

10. Boykov YY, Jolly M-P (2001) Interactive graph cuts for optimal boundary and region segmentation of objects in N-D images. In: IEEE international conference on computer vision 2001 (ICCV 2001), pp 105–112
11. Boykov YY, Jolly M-P (2001) Demonstration of segmentation with interactive graph cuts. In: IEEE international conference on computer vision 2001 (ICCV 2001), p 741
12. Kim JH, Huang RG, Jin SH et al (2009) Mobile-based flower recognition system. In: Intelligent information technology application. IITA 2009, vol 3, pp 580–583
13. Oppenheim D, Edan Y, Shani G et al (2017) Detecting tomato flowers in greenhouses using computer vision. World Acad Sci Eng Technol Int J Comput Electr Autom Control Inf Eng 11:104–109
14. Yahata S, Onishi T, Yamaguchi K et al (2017) A hybrid machine learning approach to automatic plant phenotyping for smart agriculture. In: 2017 international joint conference on neural networks (IJCNN), pp 1787–1793
15. Dias PA, Tabb A, Medeiros H (2018) Apple flower detection using deep convolutional networks. Comput Ind 99:17–28
16. Shimizu H, Hoshi T, Nakamura K et al (2015) Development of a non-contact ultrasonic pollination device. Environ Control Biol 53:85–88
17. Shimizu H, Sato T (2018) Development of strawberry pollination system using ultrasonic radiation pressure. IFAC-Papers OnLine 51:57–60
18. Lim JY, Ahn HS, Nejati M et al (2020) Deep neural network based real-time Kiwi fruit flower detection in an orchard environment
19. Shang Y, Zhang Q, Song H (2022) Application of deep learning using YOLOv5s to apple flower detection in natural scenes. Trans Chin Soc Agric Eng 38(9):222–229. https://doi.org/10.11975/j.issn.1002-6819.2022.09.024
20. Hiraguri T, Kimura T, Endo K, Ohya T, Takanashi T, Shimizu H (2023a) Shape classification technology of pollinated tomato flowers for robotic implementation. Sci Rep 13:2159. https://doi.org/10.1038/s41598-023-27971-z
21. Hong S-W, Choi L (2012) Automatic recognition of flowers through color and edge based contour detection. In: 2012 3rd international conference on image processing theory, tools and applications (IPTA), Istanbul, Turkey, pp 141–146. https://doi.org/10.1109/IPTA.2012.6469535
22. Li X, Li T, Qiu Q et al (2022) Research progress on autonomous navigation of orchard mobile robots. China J Agric Mach Chem 43(05):156–164. https://doi.org/10.13733/j.jcam.issn.2095-5553.2022.05.023
23. Hulens D, Ranst WV, Cao Y et al (2022) Autonomous visual navigation for a flower pollination drone. Machines 10(5):364. https://doi.org/10.3390/machines10050364
24. Hiraguri T, Shimizu H, Kimura T et al (2023b) Autonomous drone-based pollination system using AI classifier to replace bees for greenhouse tomato cultivation. IEEE Access 11:99352–99364. https://doi.org/10.1109/ACCESS.2023.3312151
25. Li K, Zhai L, Pan H et al (2022) Identification of the operating position and orientation of a robotic kiwifruit pollinator. Biosyst Eng 222:29–44. https://doi.org/10.1016/j.biosystemseng.2022.07.014
26. Mu X, He L, Heinemann P et al (2023) Mask R-CNN based apple flower detection and king flower identification for precision pollination. Smart Agric Technol
27. Liu L, Xu W, Yang X et al (2023) Path planning techniques for mobile robots: review and prospect. Exp Syst Appl 227:120254. https://doi.org/10.1016/j.eswa.2023.120254
28. Wei S, Li SC, Zhang M et al (2017) GNSS based automatic navigation path search and steering control for agricultural machinery. J Agric Eng 33(S1):70–77
29. Han J, Park C, Park Y et al (2019) Preliminary results of the development of a single-frequency GNSS RTK-based autonomous driving system for a speed sprayer. J Sensors 2019:1–9
30. Benedek C, Majdik A, Nagy B et al (2021) Positioning and perception in LIDAR point clouds. Digit Signal Process 119:103193
31. Jones MH, Bell J, Dredge D et al (2019) Design and testing of a heavy-duty platform for autonomous navigation in kiwifruit orchards. Biosyst Eng

32. Wu C, Li X (2023) Autonomous navigation method for orchard robots based on single line laser. J Jiangsu Univ 44(05):530–539
33. Deng Z, Li H (2016) Survey of research on mobile robot autonomous navigation technology. Sci Technol Inf 14(33):142–144
34. Radcliffe J, Cox J, Bulanon D et al (2018) Machine vision for orchard navigation. Comput Ind 98:165–171
35. Hai J, Hao Y, Zou F, Lin F et al (2021) A visual navigation system for UAV under diverse illumination conditions. Appl Artif Intell 35(15):1529–1549. https://doi.org/10.1080/08839514.2021.1985799
36. Cai Y, Qian W, Dong J et al (2023) A LiDAR-inertial SLAM method based on virtual inertial navigation system. Electronics 12:2639. https://doi.org/10.3390/electronics12122639
37. Li X, Qiu Q (2021) Autonomous navigation for orchard mobile robots: a rough review. In: 2021 36th youth academic annual conference of Chinese association of automation (YAC), Nanchang, China, pp 552–557. https://doi.org/10.1109/YAC53711.2021.9486486
38. Xie Y, Wang Y, Nallanathan A et al (2016) An improved K-nearest-neighbor in-door localization method based on spearman distance. IEEE Signal Process Lett 23:351–355
39. Sun D, Wei E, Ma Z, Wu C, Xu S (2021) Optimized CNNs to indoor localization through BLE sensors using improved PSO. Sensors 21(6):1995. https://doi.org/10.3390/s21061995
40. Ashraf I, Hur S, Park Y (2019) Indoor positioning on disparate commercial smartphones using Wi-Fi access points coverage area. Sensors 19:4351
41. Krapež P, Vidmar M, Munih M (2021) Distance measurements in UWB-radio localization systems corrected with a feedforward neural network model. Sensors 21:2294
42. Liu M, Wang H, Yang Y et al (2018) RFID 3-d indoor localization for tag and tag-free target based on interference. IEEE Trans Instrum Meas 68:3718–3732
43. Zhou R, Yang Y, Chen P (2021) An RSS transform—based WKNN for indoor positioning. Sensors 21:5685. https://doi.org/10.3390/s21175685
44. Kostavelis I, Gasteratos A (2015) Semantic mapping for mobile robotics tasks: a survey. Robotics Auton Syst 66:86–103
45. Wu W, Guo L, Gao H et al (2022) YOLO-SLAM: a semantic SLAM system towards dynamic environment with geometric constraint. Neural Comput Appl 34:6011–6026. https://doi.org/10.1007/s00521-021-06764-3
46. Wu S, Liu J, Lei X et al (2022) Research progress on efficient pollination technology of crops. Agronomy 12(11):2872. https://doi.org/10.3390/agronomy12112872
47. Sengupta S, Sturgess P, Ladický L et al (2012) Automatic dense visual semantic mapping from street-level imagery. In: 2012 IEEE/RSJ international conference on intelligent robots and systems, Vilamoura-Algarve, Portugal, pp 857–862. https://doi.org/10.1109/IROS.2012.6385958
48. Lu Y, Xue Z, Xia G et al (2018) A survey on vision-based UAV navigation. Geo-Spatial Inf Sci 21(1):21–32. https://doi.org/10.1080/10095020.2017.1420509
49. Lynen S, Achtelik MW, Weiss S et al (2013) A robust and modular multi-sensor fusion approach applied to MAV navigation. In: 2013 IEEE/RSJ international conference on intelligent robots and systems, Tokyo, Japan, pp 3923–3929. https://doi.org/10.1109/IROS.2013.6696917
50. Magree D, Johnson EN (2014) Combined laser and vision-aided inertial navigation for an indoor unmanned aerial vehicle. In: 2014 American control conference, Portland, OR, USA, pp 1900–1905. https://doi.org/10.1109/ACC.2014.6858995
51. Zhang Y, Xu X, Zhang N et al (2021) A semantic SLAM system for catadioptric panoramic cameras in dynamic environments. Sensors 21:5889. https://doi.org/10.3390/s21175889
52. Kyprianou G, Doitsidis L, Chatzichristofis SA (2021) Towards the achievement of path planning with multi-robot systems in dynamic environments. J Intell Robot Syst 104:15
53. Liu Y, Jiang C, Zhang X et al (2022) Reliability-based multivehicle path planning under uncertainty using a bio-inspired approach. J Mech Des 144:091701
54. Lin S, Liu A, Wang J et al (2022) A review of path-planning approaches for multiple mobile robots. Machines 10:773. https://doi.org/10.3390/ma-chines10090773

55. Liu L, Wang X, Yang X et al (2023) Path planning techniques for mobile robots: review and prospect. Expert Syst Appl 227:120254
56. Yang L, Li P, Qian S et al (2023) Path planning technique for mobile robots: a review. Machines 11:980. https://doi.org/10.3390/machines11100980
57. Wang Y, Su Y, Quan S et al (2022) Study on the deposition distribution of drop-lets at different spray rates during drone pollination of Korla fragrant pear. Chin Fruit Tree 12:20–25. https://doi.org/10.16626/j.cnki.issn1000-8047.2022.12.04
58. Yi X, Zhu A, Yang S et al (2016) A bio-inspired approach to task assignment of swarm robots in 3-D dynamic environments. IEEE Trans Cybern 47(4):974–983
59. Song P, Pan J, Chu S (2020) A parallel compact cuckoo search algorithm for three-dimensional path planning. Appl Soft Comput 94 (Article 106443)
60. Raz Sprayers (n.d.) Raz 500 for spraying and dusting date palm—hydraulic elevation by a boom. Raz Sprayers. https://www.razsprayers.com/product/raz-500-for-spraying-and-dusting-date-palm-hydraulic-elevation-by-a-boom/. Accessed 20 Feb 2024
61. Shapiro A, Korkidi E, Rotenberg A et al (2008) A robotic proto-type for spraying and pollinating date palm trees. In: Engineering systems design and analysis, vol 48388, pp 431–436
62. Shapiro E, Korkidi A, Demri O et al (2009) Toward elevated agrobotics: development of a scaled-down prototype for visually guided date palm tree sprayer. J Field Robot 26:572–590
63. Williams H, Bell J, Nejati M, Hussein S, Penhall N, Lim J, MacDonald B et al (2021) Evaluating the quality of kiwifruit pollinated with an autonomous robot. Field Rob 1:231–252. https://doi.org/10.55417/fr.2021009
64. Barnett J, Seabright M, Williams H et al (2017) Robotic pollination—targeting kiwifruit flowers for commercial application
65. Duke M, Barnett J, Bell J et al (2017) Automated pollination of kiwifruit flowers
66. Nejati M, Ahn H, MacDonald B (2020) Design of a sensing module for a kiwifruit flower pollinator robot. ArXiv, abs/2006.08045
67. Vaknin Y, Gan-Mor S, Bechar A, Ronen B, Eisikowitch D (2001) Electrostatic pollination of pistachio (Pistacia vera L.): a novel technique of pollen supplementation in agriculture. Cah Options Méditerranéennes 56:53–57. https://api.semanticscholar.org/CorpusID:172134099
68. Bechar A, Shmulevich I, Eisikowitch D, Vaknin Y, Ronen B, Gan-Mor S (1999) Modeling and experiment analysis of electrostatic date pollination. Trans ASABE 42:1511–1516. https://doi.org/10.13031/2013.13314
69. Bechar A, Gan-Mor S, Ronen B (2008) A method for increasing the electrostatic deposition of pollen and powder. J Electrostat 66:375–380
70. Shaneyfelt T, Jamshidi M, Agaian S (2013) A vision feedback robotic docking crane system with application to vanilla pollination. Int J Autom Control 7:62–82
71. Yu X, Kong DY, Xie X et al (2022) Target recognition and detection of tomato pollination robot based on deep learning. Trans Chin Soc Agric Eng 38(24):129–137
72. Barnett J, Seabright M, Williams H et al (2017) Robotic pollination—targeting kiwifruit flowers for commercial application (2017)
73. Wood R, Nagpal R, Wei G (2013) Flight of the robobees. Sci Am 308(3):60–65. http://www.jstor.org/stable/26018027
74. Kevin YM, Pakpong C, Sawyer BF et al (2013) Controlled flight of a biologically inspired, insect-scale robot. Science 340:603–607. https://doi.org/10.1126/science.1231806
75. Savage N (2015) Aerodynamics: vortices and robobees. Nature 521(7552):S64–S65
76. Berman S, Kumar V, Nagpal R (2011) Design of control policies for spatially inhomogeneous robot swarms with application to commercial pollination. In: 2011 IEEE international conference on robotics and automation, Shanghai, China, pp 378–385. https://doi.org/10.1109/ICRA.2011.5980440
77. Berman S, Nagpal R, Halász Á (2011) Optimization of stochastic strategies for spatially inhomogeneous robot swarms: a case study in commercial pollination. In: 2011 IEEE/RSJ international conference on intelligent robots and systems, San Francisco, CA, USA, pp 3923–3930. https://doi.org/10.1109/IROS.2011.6094771

78. Elamvazhuthi K, Berman S (2015) Optimal control of stochastic coverage strategies for robotic swarms. In: 2015 IEEE international conference on robotics and automation (ICRA), Seattle, WA, USA, pp 1822–1829. https://doi.org/10.1109/ICRA.2015.7139435
79. Abutalipov RN, Bolgov YV, Senov HM (2016) Flowering plants pollination robotic system for greenhouses by means of nano copter (drone aircraft). In: 2016 IEEE international conference on quality management, transport and information security, information technologies. IEEE 2016, pp 7–9
80. Futurism (n.d.) Autonomous robot bees are being patented by...Walmart? https://www.cnet.com/science/walmart-to-make-autonomous-ro-bot-bees-pollen-drones/
81. CanNews (n.d.) https://ep.cannews.com.cn/publish/zghkb7/-html/4905//node_204341.html
82. Yang J, Zhang H, Berdin A, Hu W, Zeng H (2023) Dandelion-inspired, wind-dispersed polymer assembly controlled by light. Adv Sci 10(7):2206752
83. Priscilla J, Roshni T, Edwin B (2022) Smart self-pollination for future agricultural—a computational structure for micro air vehicles with man-made and artificial intelligence. Int J Intell Syst Appl Eng 10:170–174
84. Potts SG, Neumann P, Vaissière B et al (2018) Robotic bees for crop pollination: Why drones cannot replace biodiversity. Sci Total Environ 642:665–667. https://doi.org/10.1016/j.scitotenv.2018.06.114. Epub 2018 Jun 14. PMID: 29909334
85. Praveen V, Pillai S (2016) Modeling and simulation of quadcopter using PID controller. Int J Control Theory Appl 9(15):7151–7158
86. Dropcopter. (n.d.) Agricultural aerial solutions. https://www.dropcopter.com/agricultural-aerial-solutions
87. Chechetka S, Yu Y, Tange EM et al (2017) Materially engineered artificial pollinators. Chem 2:224–239
88. Yang X, Miyako E (2020) Soap bubble pollination. iScience 23(6):101188. https://doi.org/10.1016/j.isci.2020.101188. Epub 2020 Jun 17. PMID: 32553132; PMCID: PMC7327714
89. Hulens D, Ranst WV, Cao Y et al (2022) Autonomous visual navigation for a flower pollination drone. Machines 10(5):364. https://doi.org/10.3390/machines10050364.
90. Mazinani M, Dehghani M, Zarafshan P et al (2021) Design and manufacture of an aerial pollinator robot for walnut trees. In: 2021 9th RSI international conference on robotics and mechatronics (ICRoM), Tehran, Islamic Republic of Iran, pp 445–450. https://doi.org/10.1109/ICRoM54204.2021.9663500
91. Mazinani M, Zarafshan P, Dehghani M et al (2023) Design and analysis of an aerial pollination system for walnut trees. Biosys Eng 225:83–98
92. Rice CR, McDonald ST, Shi Y et al (2022) Perception, path planning, and flight control for a drone-enabled autonomous pollination system. Robotics 11(6):144. https://doi.org/10.3390/robotics11060144
93. Wang S, Lei X, Tang Y et al (2020) Pear spray pollination technology based on multi rotor UAV. Jiangsu Agric Sci 48(23):210–214. https://doi.org/10.15889/j.issn.1002-1302.2020.23.043
94. Wang Y, Bai R, Lu X et al (2022) Pollination parameter optimization and field verification of UAV-based pollination of 'Kuerle Xiangli'. Agronomy 12(10):2561. https://doi.org/10.3390/agronomy12102561
95. Jia Y, Wang X, Wang R et al (2023) Simulation study on multi rotor unmanned aerial vehicle (UAV) orchard cluster pollination. Agric Mech Res 45(06):24–29. https://doi.org/10.13427/j.cnki.njyi.2023.06.010
96. Shi Q, Liu D, Mao H et al (2019) Study on assistant pollination of facility tomato by UAV. 2019 Boston, Massachusetts (July 7–July 10, 2019)
97. Yao F (2022) Development of pneumatic pollination components for large-scale hybrid rice seed production (Doctoral dissertation, Zhejiang University). https://doi.org/10.27461/d.cnki.gzjdx.2020.004116
98. Su H (2020) Study on the effect of different mechanical assisted pollination methods on hybrid rice seed production (Master's thesis, Hunan Agricultural University). https://doi.org/10.27136/d.cnki.ghunu.2019.000217

99. Li Z, Tang C, Li M et al (2015) Optimization and experimental verification of parameters of pneumatic pollination jet pipeline. Trans Chin Soc Agric Eng 31(21):68–75
100. Li Z, Tang C, Wang H, Li M, Huang Z, Fang S (2012) Effects of location of airflow on pollen distribution for pneumatic pollination in hybrid rice breeding. Trans Chin Soc Agric Eng 28(1):107–113. https://doi.org/10.3969/j.issn.1002-6819.2012.z2.019
101. Wu L, Fang S, Li Z et al (2012) Collision pneumatic-type hybrid-rice seed-production pollination machine
102. Weng X, Xu J, Zhao J et al (2022) Research on the assistance of multi-rotor UAVs in hybrid rice seed production for Indica and Japonica varieties. J Chin Agric Mech 43(09):24–29. https://doi.org/10.13733/j.jcam.issn.2095-5553.2022.09.004
103. Li J, Lan Y, Wang J et al (2017) Int J Agric Biol Eng 10(4):32–40. https://doi.org/10.25165/j.ijabe.20171004.3103
104. Wu H, Xiong C, Liu A et al (2014) Research status and ideas on mechanized seed production and assisted pollination technology for hybrid rice. Crop Res 28(03):321–323+327
105. Wang P, Hu L, Zhou Z et al (2013) Field wind field measurement of unmanned oil powered helicopters for assisted pollination in rice seed production. J Agric Eng 29(03):54–61+294
106. Li J, Zhou Z, Lan Y et al (2015) Distribution pattern of canopy wind field during rotary wing unmanned aerial vehicle pollination operation. J Agric Eng 31(03):77–86
107. Shi Q, Liu D, Mao H (2021) Wind-induced response of rice under the action of the downwash flow field of a multi-rotor UAV. Biosys Eng 203:60–69
108. Feng L, Wu W, Wang J et al (2019) Wind field distribution of multi-rotor UAV and its influence on spectral information acquisition of rice canopies. Remote Sensing 11(6):602. https://doi.org/10.3390/rs11060602
109. Wang J, Liu J, Chen W et al (2022) Robot path planning via neural-network-driven prediction. IEEE Trans Artif Intell 3(3):451–460. https://doi.org/10.1109/TAI.2021.3119890

Lightweight and Robust YOLOv5s with MobileNetV3 and GhostNet for Precision Apple Flower Detection for Pollination Drones

Shahram Hamza Manzoor, Zhao Zhang, Xu Li, and Liling Yang

Abstract The improved YOLOv5s model represents a meticulous endeavor to balance efficiency and accuracy in apple flower detection within natural environments for pollination drones. Through strategic modifications, leveraging a MobileNetV3 backbone and GhostNet head, the model achieves a significant reduction in computational complexity while maintaining commendable performance metrics. Comparative analysis positions it as a standout choice, bridging the accuracy gap between modified YOLOv5s with MobileNetV3 backbone and GhostNet head and YOLOv5s, YOLOv7 and YOLOv8s while outperforming alternatives like Faster R-CNN. Precision, recall, and mean Average Precision (mAP50) metrics underscore its efficacy, with precision and recall values reaching 90.6% and 87.7%, respectively, alongside a remarkable mAP50 score of 91.2%. Crucially, the reduction in GFLOPs and model size by approximately 89% and 85%, respectively as compared to YOLOv5s benchmark model, enhancing its deploy ability on constrained devices. Execution tests on embedded devices further validate its practicality, demonstrating impressive 227 FPS on NVIDIA GeForce RTX 4060 Ti GPU device and 20 Frames Per Second (FPS) ratings, particularly on resource-constrained platforms like the Jetson devices. Looking ahead, opportunities for fine-tuning under specific environmental conditions and extending applicability to diverse flora remain, alongside the exploration of edge computing optimizations to further streamline deployment. This modified YOLOv5s emerges as a robust solution, poised to address the demand for lightweight yet accurate computer vision models in real-world scenarios for pollination dronesS.

S. H. Manzoor · Z. Zhang (✉)
Key Laboratory of Smart Agriculture System Integration, Ministry of Education, China Agricultural University, Beijing 100083, China
e-mail: zhaozhangcau@cau.edu.cn

Key Laboratory of Agricultural Information Acquisition Technology, Ministry of Agriculture and Rural Affairs, China Agricultural University, Beijing 100083, China

X. Li
College of Information Engineering, Tarim University, Aral 843300, Xinjiang, China

L. Yang
Institute of Agricultural Mechanization, Xinjiang Academy of Agricultural Sciences, Ürümqi 830091, Xinjiang, China

Y. Liu et al. (eds.), *New Technologies Applied in Apple Production*, Smart Agriculture 10, https://doi.org/10.1007/978-981-97-7778-5_4

Keywords MobileNetV3 · GhostNet · YOLOv5 · YOLOv6 · YOLOv8 · R-CNN · GFLOPs

1 Introduction

In the ever-evolving landscape of global agriculture, the indispensable role of apple cultivation cannot be overstated [1]. As the world's leading apple producer, China has apple production is 45 million metric tons (MMT) for marketing year 2023/24 as shown in Chart 1. China faces challenges that threaten the stability of its apple production, with adverse weather conditions during pollination significantly impacting provincial apple yields [2]. Apples, essential for their nutritional benefits, heavily depend on efficient cross-pollination [3, 4]. However, traditional pollinators such as bees and insects are dwindling due to factors such as climate change [3]. This necessitates innovative solutions to ensure the stability of apple production.

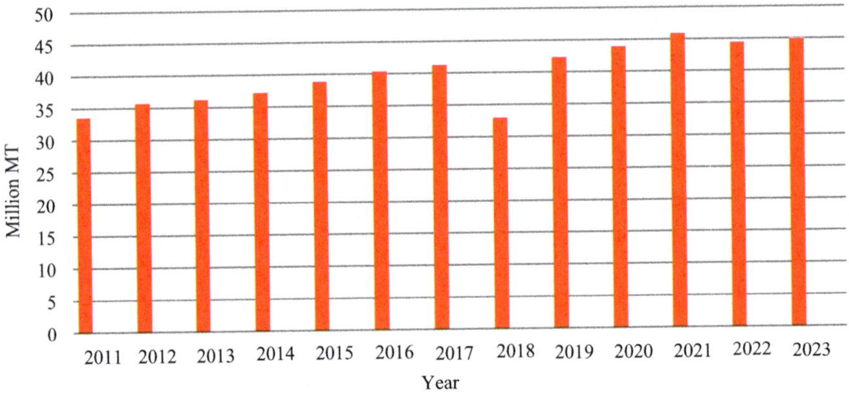

Chart 1 China: apple production. *Source* National Bureau of Statistics (NBS), FAS Beijing [2]

This study introduces a pioneering approach by focusing on the integration of a modified YOLOv5, a state-of-the-art object detection model with edge devices for apple flower pollination drone technology. Tailored specifically for robust apple flower detection, this synergistic methodology stands poised to confront the challenges of pollination, presenting a precise solution for apple cultivation.

Amid global concerns about the decline of traditional pollinators, this research underscores the pivotal role of YOLOv5 in significantly enhancing the efficiency of drone pollination methods. The paper unfolds the untapped potential of YOLOv5 in addressing challenges associated with traditional pollination, providing an advanced solution for apple flower detection. The discussion culminates in highlighting the transformative impact of this modified YOLOv5 model on apple flower pollination, promising not only enhanced yields but also paving the way for a sustainable and technologically advanced future in apple cultivation.

1.1 Related Literature for Apple Flower Detection

The exploration of apple flower detection for real-time applications, especially in the context of pollination, is a critical yet underexplored domain in agricultural research. Numerous studies have made significant contributions to identifying and quantifying apple flowers, but a predominant focus on achieving enhanced detection accuracy has often overshadowed the crucial aspect of real-time applicability.

Several studies have carved unique niches in the intricate domain of apple flower detection. Dias et al. [5], utilizing a pre-trained convolutional neural network (CNN), exhibits resilience to environmental factors but unintentionally prioritizes accuracy over the real-time imperative for practical scenarios like precision pollination. Detection results are shown in Fig. 1. Zhang et al. [6], involving a fusion of unmanned aerial vehicles (UAV) and ground-based RGB image data, delves into flowering intensity estimation as shown in Fig. 2. However, its primary focus on automated thinning guidance inadvertently sidesteps the potential for real-time applications in precision pollination. Bhattarai and Karkee [7] introduces a promising weakly-supervised flower counting network (CountNet) as shown in Fig. 3, but the absence of explicit exploration into real-time deployment and applications beyond counting underscores a broader research gap. Wang et al. [8] emphasis on accurate flower density mapping for variable-rate chemical sprayers is noteworthy, but its preoccupation with segmentation accuracy eclipses the critical need for real-time interventions like precision pollination. Results of this study are presented in Fig. 4. Figure 5 shows the trajectory of real-time apple flower detection takes center stage in [9], employing the YOLOv4 deep learning algorithm. However, despite acknowledging the feasibility of real-time detection, the primary application remains confined to flower detection, missing a deeper exploration of interventions such as precision pollination. Shang et al. [10], using the modified YOLOv5s algorithm, leans towards orchard flower thinning rather than the broader spectrum of precision pollination scenarios. Detection results are given in Fig. 6.

Tian et al. [11] introduces an instance segmentation model for apple flower detection shown in Fig. 7, but falters in providing insights into real-time applications or embedded device deployment. While [12] may not be directly tied to apple flower

Fig. 1 Examples of images composing the additional datasets Apple B (left), Apple C (middle) and Peach (right), overlaid with the corresponding detections obtained by this study [5]

Fig. 2 Left: coloured point cloud of tree. Middle: section of tree with clear flower blossoms present. Right: flower clusters detected from the middle volume using the Otsu segmentation method [6]

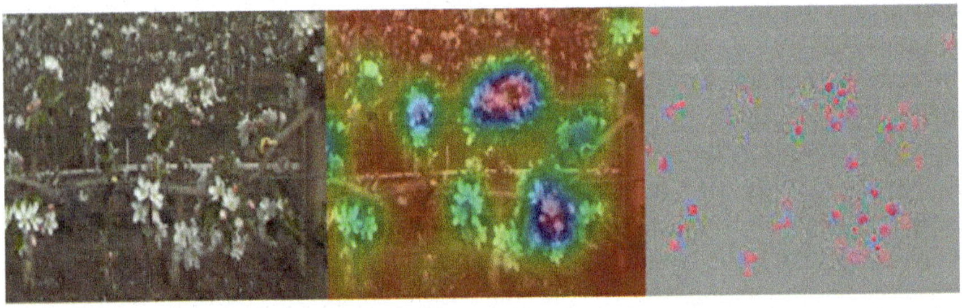

Fig. 3 Detection results of study [7]

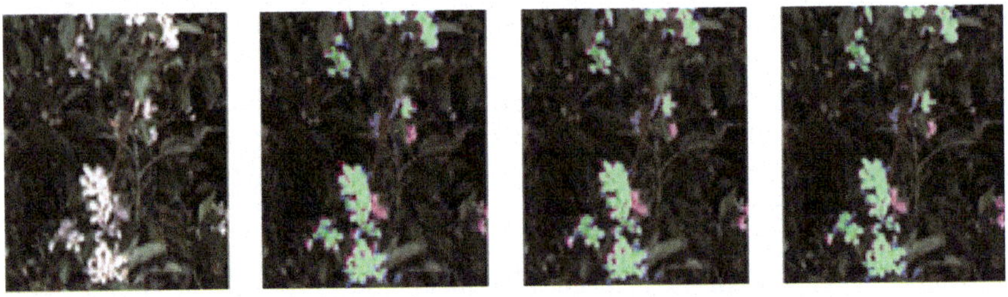

Fig. 4 FCNs-Edge results on the same image with different input resolutions [8]

Fig. 5 Apple flower detection results based on YOLO v4 algorithm. **a**–**c** are Fuji, Red Love and Gala, respectively [9]

Fig. 6 Detection effect of four different weather of flowers: **a** detection effect of sunny; **b** detection effect of cloudy; **c** detection effect of overcast sky and **d** detection effect of light rainy [10]

pollination, its rapid pathogen detection protocol using loop-mediated isothermal amplification (LAMP) underscores the importance of timely interventions, indirectly emphasizing the significance of real-time applications in related fields. Mu et al. [13] focus on identifying individual king flowers within apple flower clusters brings attention to the importance of monitoring blooming stages presented in Fig. 8. Although the study refrains from explicit emphasis on real-time applications, it leaves room for further exploration in the context of precision pollination. Yuan [14] weakly-supervised flower counting network strives to simplify automated counting methods but stops short of explicitly addressing real-time deployment or applications beyond counting.

Chen [15] orchard blooming level estimation method, enriched with an improved YOLOv5-based apple flower detector with relatively lower accuracy, holds potential for orchard management but neglects explicit exploration of real-time applications or interventions like precision pollination. Detection results are shown in Fig. 9. Zhang et al. [16], addressing network model stability, introduces an optimized detection network model. Yet, its primary application remains confined to apple flower detection, lacking explicit focus on real-time interventions. Dias et al. [17] introduces an automated technique for flower identification using an end-to-end residual convolutional neural network (CNN) with robustness across different flower

Fig. 7 Segmentation results of MASU ReCNN and Mask ReCNN [11]

Fig. 8 Detection of King flowers [13]

species including apple flowers as shown in Fig. 10. However, it does not explicitly address real-time applications. Hočevar et al. [18], focusing on tree-specific practices in high-density apple orchards, employs image analysis to estimate the number of flower clusters shown in Fig. 11. Its reluctance to delve into real-time applications or interventions like precision pollination leaves a void.

Farjon [19] study introduces system for estimating blooming intensity and peak blooming dates presents potential applications for predicting tree yield and blooming dates but leans towards orchard management rather than real-time applications like precision pollination. Yuan et al. [20] quantification of the sensitivity of YOLOv4 to artificial image distortions lacks explicit exploration of real-time applications or interventions. Aggelopoulou et al. [21] introduces Deep Phenology, a novel phenology distribution estimation method based on CNNs for apple flowers, striving to efficiently map flower distribution on different levels but omits explicit focus on real-time applications like precision pollination.

While all of these studies significantly advance the accuracy and efficiency of apple flower detection methods, a conspicuous gap emerges. The majority prioritize

Fig. 9 Comparison of the detection results of the model in this paper and the YOLOv5s model for apple flowers at the 'Petal Fall' stage. **a** Our model and **b** YOLOv5s [15]

Fig. 10 Flower detection in one image composing the Apple dataset [17]

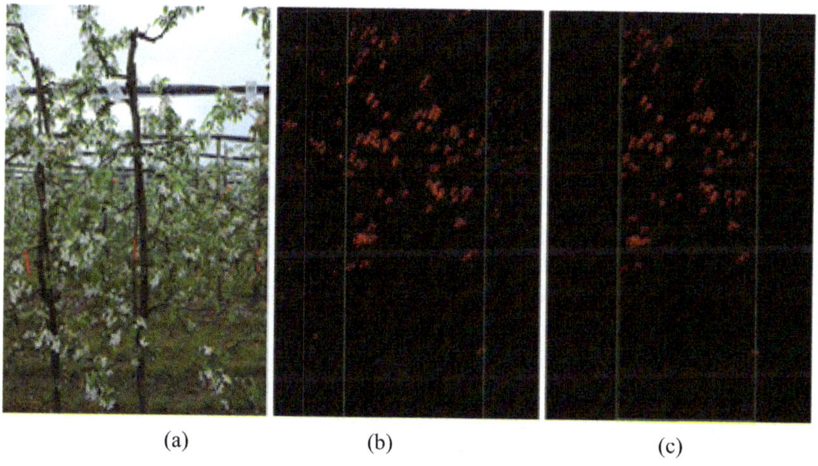

Fig. 11 Example of image processing taken during the day with still color camera. **a** Original image, **b** image after HSL thresholding and **c** image after thresholding and rejection of objects outside the tree frame (green) and rejection of too large and too small objects [18]

detection capabilities, inadvertently sidelining the critical need to explore real-time applications for interventions like precision pollination. This introduction sets the stage for a meticulous exploration of this research gap, offering a foundation for the imperative deployment of apple flower detection in real-time precision agriculture. Some of these studies contributing uniquely to the evolution of YOLOv5. By combining insights from seemingly unrelated domains, this narrative paints a vivid picture of YOLOv5's adaptability, versatility, and its ability to transcend boundaries, providing a comprehensive guide for specialized object detection, with a focus on the nuanced challenges of apple flower detection for pollination drones in orchards., the fusion of YOLOv5 and drone technology emerges as a beacon of hope for addressing the challenges faced by China's apple production.

1.2 Research Objective and Significance

This research advances precision agriculture by enhancing YOLOv5s for precise apple flower detection in pollination drones. The focus is on optimizing YOLOv5s for real-time deployment on embedded devices like Jetson and Raspberry Pi, overcoming computational hurdles. Key modifications include integrating MobileNetV3 into YOLOv5s backbone for improved efficiency and preserving robust feature extraction tailored to apple flower nuances. Additionally, GhostNet in the head module boosts efficiency and collaborates with MobileNetV3 for nuanced feature enhancement.

2 Materials and Method

2.1 Technical Specifications

The technical specifications employed in this study facilitated the efficient execution of our research tasks. Our operating system of choice was Linux-6.5.0-14-generic-x86_64-with-glibc2.17. Python version 3.8.18 served as the primary programming language for algorithm development and implementation. To harness accelerated computation capabilities, we leveraged a NVIDIA GeForce RTX 4060 Ti GPU device boasting a total memory capacity of 8 GB. These specifications provided the necessary computational power to proficiently conduct image processing, model training, and evaluation tasks pivotal to our research objectives. We conducted execution tests on embedded devices, focusing on the practical deployment of the Modified YOLOv5 on Jetson Xavier NX. This further extended our investigation into real-world deployment scenarios, ensuring the robustness and applicability of our proposed methodologies.

2.2 Data Collection

During the meticulous dataset preparation phase, 3416 images of apple flowers were meticulously captured using a HUAWEI Mate 50 camera, each image possessing a resolution of 4096 × 3072 pixels. These images underwent rigorous processing to ensure uniformity and optimal conditions for subsequent model training and evaluation. The dataset was partitioned such that 80% of the images were allocated for training, 10% for testing, and 10% for validation purposes, ensuring a balanced distribution across the dataset subsets. To mitigate potential variations introduced by diverse camera settings, each image underwent auto-orientation with EXIF metadata stripping. Subsequently, employing a stretch operation, all images were resized to a standardized dimension of 640 $\times$ 640 pixels while maintaining aspect ratio consistency.

2.3 *Data Augmentation*

The integration of MobileNetV3 for feature extraction was complemented by a robust set of data augmentation techniques, including horizontal flipping (50% probability), random rotation (-15 to $+15°$) for handling varied flower orientations, and random shear ($-15°$ to $+15°$ horizontally and vertically) to address non-ideal camera angles. Moreover, random adjustments for brightness and exposure (-25 to $+25\%$) tackled challenges posed by diverse lighting conditions. The application of salt and pepper noise (5% of pixels) simulated real-world artifacts, fortifying the model's resilience to noisy images. All the augmentation techniques applied are shown in Fig. 12. This comprehensive approach ensures the model's robustness, critical for recognizing apple flowers with varying colors and under diverse lighting conditions. The dataset encompasses images featuring primary colors (white, pink, and others corresponding to common apple varieties in China), enabling the development of a versatile detection model. Additionally, the dataset deliberately includes images captured in both front-light and back-light conditions to account for variations in brightness. Challenges arising from the coexistence of flowers and leaves, as well as the dense distribution of flowers within clusters, were considered, acknowledging potential obstructions to flower visibility. The careful curation of this dataset fosters the model's adaptability and generalization capabilities, laying a solid foundation for improved performance during subsequent training and evaluation stages.

2.4 *YOLOv5 Architectures and Its Limitations*

The evolution of deep learning in object detection is exemplified by YOLOv5, a groundbreaking model developed by Ultralytics [22–24]. Originally designed for a diverse range of applications, YOLOv5 has been meticulously fine-tuned to meet the specific requirements of detecting apple flowers for pollination drones. This exploration involves a detailed analysis of both the original YOLOv5 architecture and its modified version, highlighting their strengths, shortcomings, and how the adjustments align with the demands of efficient and precise apple flower detection.

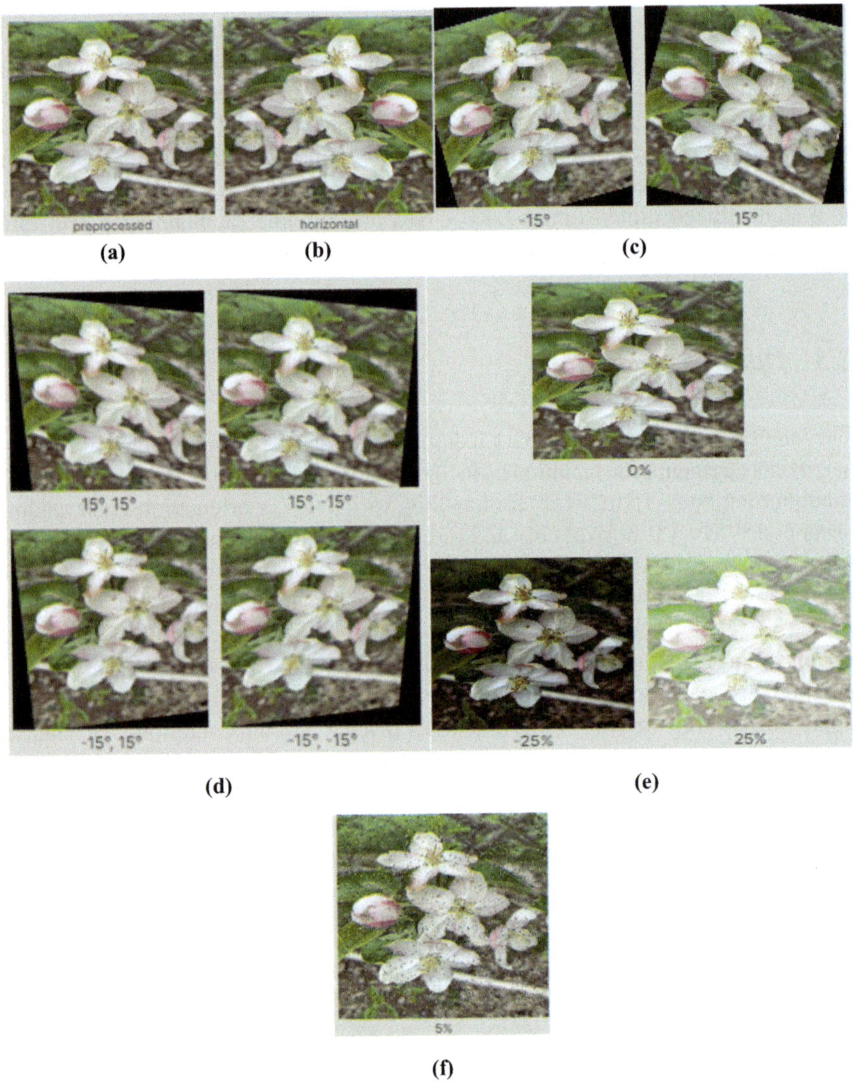

Fig. 12 Example of Apple flower images augmentation processes **a** Preprocessed image, **b** flipped image, **c** random rotated images, **d** random sheared images, **e** brightness and exposure adjusted images and **f** salt and pepper noised image

The YOLOv5 model is structured around a unified framework encompassing input, backbone, neck, and prediction components. Its versatility is showcased through variants like YOLOv5s, YOLOv5m, YOLOv5l, and YOLOv5x, each tailored to different computational resources [22, 23, 25]. This design flexibility allows the model to seamlessly adapt to various applications, from real-time object detection to specialized tasks.

The backbone of YOLOv5s incorporates focus, Convolution (Conv), CSP bottleneck with 3 × 3 Convolution (C3), and Spatial pyramid pooling (SPP) modules as shown in Fig. 13 [22–26]. The focus module, employing innovative downsampling techniques, lays the foundation for subsequent feature extraction. The Conv unit represents the basic convolutional operation, while the C3 module introduces bottleneck residuals for efficient feature transfer. The SPP module enriches the network by executing maximum pooling with diverse kernel sizes, enhancing feature diversity [27, 28].

While YOLOv5's general-purpose nature showcases its versatility, the computational intensity of the backbone design can be a limiting factor when deploying the model on embedded devices. The intricate design of the backbone may demand significant computational resources, potentially hindering its real-time applicability on devices with constrained processing capabilities.

In its original design, YOLOv5 is crafted to cater to a broad spectrum of object detection scenarios. However, this generality may lead to suboptimal performance for specific applications, such as the nuanced detection of apple flowers for pollination drones. The subsequent analysis will delve deeper into the modifications made to YOLOv5s to address these specific challenges and optimize its capabilities for efficient and precise apple flower detection in the context of pollination drones.

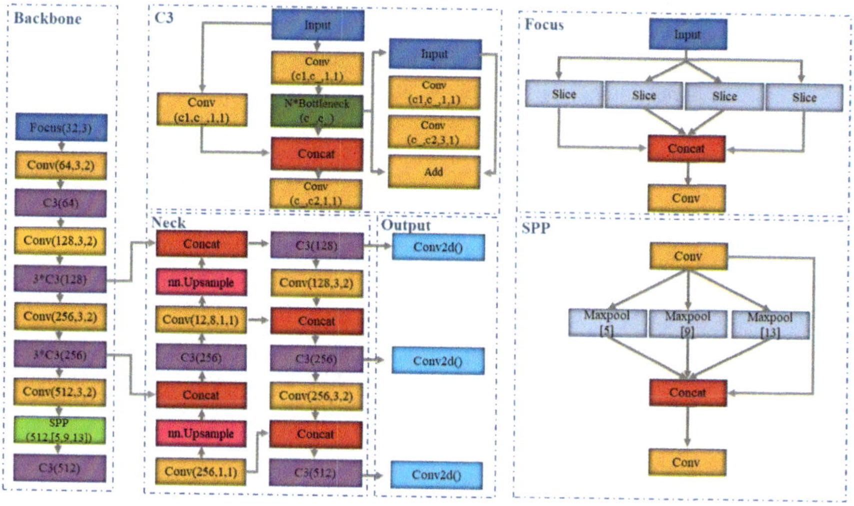

Fig. 13 YOLOv5 Architectures [22–28]

Table 1 Lightweight network performance comparison on cifar10 [22]

Network	FLOPs (G)	Memory (MB)	Latency (ms)
MobileNetV3 small 1.0 ×	0.068	18.99	27
ShuffleNetV2 1.0 ×	0.149	20.84	41
GhostNet 1.0 ×	0.150	40.04	90

2.5 *Modified YOLOv5s Architecture*

2.5.1 Improved Backbone: MobileNetV3 Integration

In 2019, Howard in [29] introduced MobileNet version 3(MobileNetV3) networks, the latest iteration in the MobileNet series renowned for its compact parameters, high accuracy, and swift real-time detection capabilities [22, 30]. Widely adopted in embedded and mobile terminals, MobileNetV3 is crafted from multiple basic Block modules, with v3-Large and v3-Small structures achieved through stacking different numbers of parameterized block modules. MobileNetV3-Large encompasses 15 Block modules, while MobileNetV3-Small has 11, featuring fewer channels [29, 30]. The MobileNetV3-Small structure, illustrated in Fig. 17, serves as the backbone for the modified YOLOv5s in this study, tailored for apple flower detection for pollination drones. Notably, the feature map in MobileNetV3 undergoes a 5 × scale, resulting in the last convolutional layer's output being only 1/5 of the size of the input image. This strategic utilization of MobileNetV3-Small as the backbone ensures efficient feature extraction and aligns with the specific demands of apple flower detection in the improved YOLOv5s architecture [22]. The study demonstrates that MobileNetV3 surpasses competitors such as ShuffleNetV2 [10] and GhostNet [31] in various metrics including computational efficiency, memory consumption, and latency [22]. The inference latency comparison was conducted on a Raspberry Pi 4b platform utilizing the NCNN framework, as illustrated in Table 1 of the study.

This integration is especially beneficial for apple flower detection, where the lightweight MobileNetV3 backbone ensures efficient feature extraction and accurate identification. The model excels in resource optimization while maintaining high performance.

2.5.2 Depthwise Separable Convolutions for Computational Efficiency

In the MobileNetV3-Small model, the 11 Blocks have two different structures, as shown in Figs. 14 and 15. From Fig. 14, it can be found that the feature map first goes through a 1 × 1 convolution, which compresses the number of channels in the feature map. Then, a [32] depth-wise separable convolution is applied to the feature map to reduce the computational cost, where n is 3 or 5. In traditional convolution, each convolutional kernel will convolve each channel of the feature map once, while in depth-wise separable convolution, each convolutional kernel will convolve only a single channel of the feature map, greatly reducing the computational cost. After the depth-wise separable convolution operation, the feature map is connected to a 1 × 1 convolution, which specifies the number of channels and aggregates the feature map. Finally, the output feature maps of the two paths are added together to obtain the final feature map. The structure shown in Fig. 14 is used in second and third Block of the MobileNetV3-Small module, and the remaining Block uses the Structure2 as shown in Fig. 15. Comparing Fig. 14 with Fig. 15, the SE module [33] is introduced in Fig. 15, which is to enhance the semantic information of the target area. In the realm of real-time detection, computational efficiency becomes paramount, and MobileNetV3 addresses this demand with unparalleled precision. The ingenious use of depth-wise separable convolutions underscores MobileNetV3's commitment to computational efficiency. In the context of apple flower detection, this efficiency is not merely a desirable feature but a crucial one, enabling the model to swiftly analyze images and provide timely results. The judicious application of depth-wise separable convolutions optimizes the computational load without compromising the model's accuracy, making it well-suited for real-time applications.

Table 2 is a detailed comparing the calculations between Standard Convolution and Depth-wise Separable Convolution (DSC) for a 6 × 6 feature map with three channels:

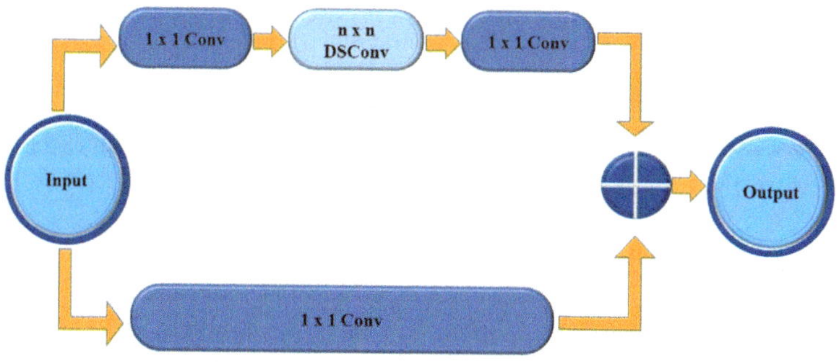

Fig. 14 Structure 1

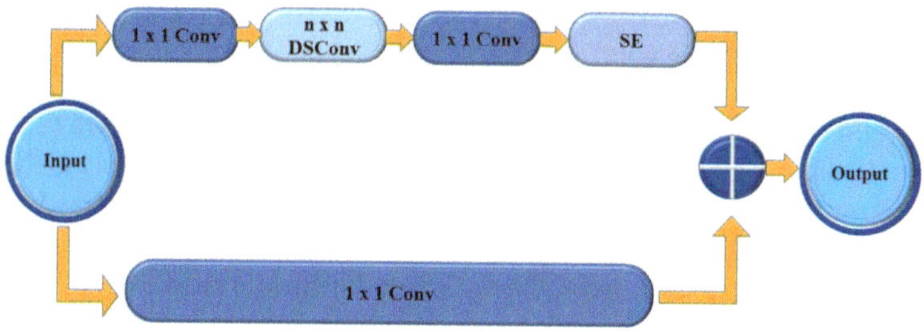

Fig. 15 Structure 2

Table 2 Comparison of calculations between standard convolution and depth-wise separable convolution (DSC)

Criteria	Standard convolution	Depth-wise separable convolution
Number of parameters	$N_{oc} + (K^2 \times C_{in} \times C_{out})$	$N_{dsc} = (K^2 \times C_{in} + C_{in} \times C_{out})$
Parameters calculation	$N_{oc} = (3^2 \times 3 \times 3) = 81$	$N_{dsc} = (3^2 \times 3 + 3 \times 3) = 36$
Computation	$C_{oc} = (K^2 \times C_{in} \times C_{out} \times H \times W)$	$C_{dsc} = (K^2 \times C_{in} \times H \times W + C_{in} \times C_{out} \times H \times W)$
Computation calculation	$C_{oc} = (3^2 \times 3 \times 3 \times 6 \times 6) = 2916$	$C_{dsc} = (3^2 \times 3 \times 6 \times 6 + 3 \times 3 \times 6 \times 6) = 1296$

Here, K is the kernel size, C_{in} is the number of input channels, C_{out} is the number of output channels, H is the height, and W is the width of the feature map. Depth-wise Separable Convolution (DSC) reduced the computational cost upto 55% as compared to Standard Convolution.

MobileNetV3 introduces a lightweight design to the backbone, striking a harmonious balance between computational efficiency and robust feature extraction. This integration ensures the model's adaptability to the resource constraints typically encountered in edge devices like Jetson or Raspberry Pi. The optimization achieved through MobileNetV3 aligns the modified YOLOv5s model with the specific demands of edge devices, guaranteeing seamless real-time operation a critical aspect for applications in edge computing environments.

2.5.3 H-Swish Activation Function and Squeeze-and-Excitation (SE) Modules of MobilenetV3

The incorporation of the h-swish activation function and the [33] Squeeze-and-Excitation (SE) module in MobileNetV3 significantly enhances efficiency for apple flower detection. The h-swish function, an evolution of swish, addresses gradient-related challenges by replacing the sigmoid function with ReLU6(x + 3)/6 [29]. This modification not only reduces computational complexity but also improves overall model performance. The Squeeze-and-Excitation (SE) attention mechanism involves two sequential steps to enhance feature representation. First, the Squeeze step employs global average pooling (GAP) on the input feature map X with dimensions $H \times W \times C$, where C represents the number of feature channels, and $H \times W$ is the height and width of the feature map. This operation compresses the feature map, generating a globally representative feature vector. Subsequently, the Excitation step utilizes two fully connected layers (Fc) to derive normalized weights. The process involves introducing learnable weights through Eq. 1, capturing intricate relationships among features. The rectified linear unit (ReLU) is then applied to this output, preserving positive values and adding a non-linear component to the operation. The following element-wise multiplication of Eq. 2, dynamically scales the original features based on the learned relationships, emphasizing relevant information. Finally, the h-swish activation function is applied to the entire expression, resulting in Eq. 3. Sequence of whole process is shown in Fig. 16.

$$\mathrm{Fc(GAP(X))} \tag{1}$$

$$\mathrm{Fc(ReLU(Fc(GAP(X))) \cdot X)} \tag{2}$$

$$\mathrm{Y = h - swish(Fc(ReLU(Fc(GAP(X)) \cdot X)))} \tag{3}$$

This application of h-swish introduces non-linearity and ensures a smooth gradient flow in the overall operation, contributing to the precision of the apple flower detection model while reducing computational complexity.

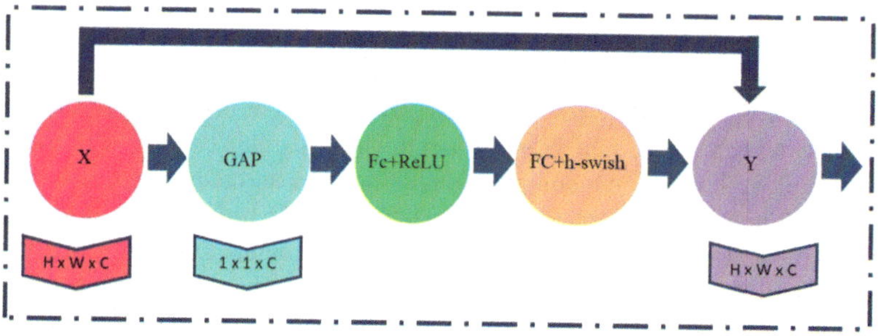

Fig. 16 Basic composition of MobilenetV3's Squeeze and excite (SE) module

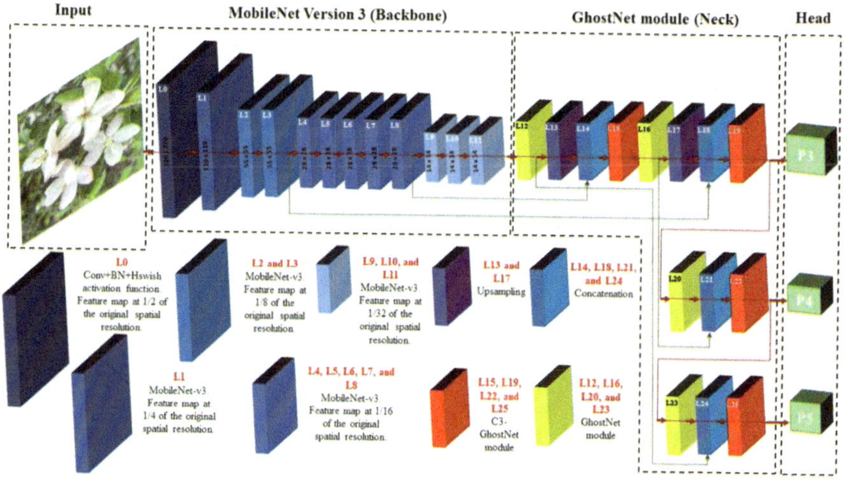

Fig. 17 Improved Yolov5s architecture

2.5.4 Attention Mechanism for Relevant Features

Embedded in Squeeze-and-Excitation SE blocks, the attention mechanism ensures that MobileNetV3 prioritizes relevant features during the detection process. This precision in feature selection is paramount for apple flowers, as it guarantees that the model concentrates on the specific visual cues' indicative of their presence. The attention mechanism augments the model's ability to discern intricate details, contributing significantly to its overall precision.

2.5.5 In-depth Analysis of MobileNetV3 Backbone Layers for Apple Flowers Detection

The MobileNetV3 architecture comprises layers designed for efficient and precise feature extraction in image processing, particularly for recognizing apple flowers for pollination drones as shown in Fig. 17. The initial layer utilizes a Convolutional Layer (Conv) with batch normalization (BN) and H-swish operation, employing 16 channels and a stride of 2 to capture low-level features like petal edges shown in Fig. 17. Subsequent layers introduce MobileNetV3 blocks, incrementing in complexity and channel count. Notably, depthwise separable convolutions are strategically employed, enhancing computational efficiency and reducing model parameters. Layer 3 maintains spatial resolution with a stride of 1, utilizing a MobileNetV3 block with 24 initial channels and 88 subsequent channels to capture fine-grained details. Layer 4 introduces a larger kernel size (5×5) and a stride of 2, preparing the model for higher-level abstraction with 40 initial channels and 96 subsequent channels. Depthwise separable convolutions continue to contribute to efficiency. Layer 5 emphasizes richer feature representations with 240 output channels, a 5×5 kernel, and a stride of 1, optimizing adaptability to diverse apple flower variations. Layers 6–8 further refine the model's understanding of intricate details specific to apple flowers. The inclusion of padding in Layer 6 (1, 1) preserves spatial information, and depthwise separable convolutions maintain computational efficiency. Layer 9 incorporates the Squeeze-and-Excitation (SE) mechanism, enhancing discriminative power with 96 initial channels and 288 subsequent channels. The depthwise separable convolution combined with SE ensures both efficiency and improved feature representation. Layers 10 and 11 employ MobileNetV3 blocks with attention mechanisms, prioritizing relevant features during detection. With parameters [96, 576, 5, 1, 1, 1], these layers significantly contribute to overall precision by focusing on specific visual cues indicative of apple flowers. In summary, MobileNetV3's architecture meticulously orchestrates convolutional layers, MobileNetV3 blocks, and depthwise separable convolutions to balance efficiency and accuracy in capturing intricate details of apple flower images.

MobileNetV3 excels in efficiency and precision for apple flower detection in YOLOv5. Its tailored design, featuring hierarchical feature pyramids, depthwise separable convolutions, and adaptive SE blocks, makes it a powerhouse. The model efficiently handles variations in apple flower characteristics, ensuring real-time detection on resource-constrained devices. MobileNetV3's seamless alignment with detection requirements, combined with purposeful data augmentations, demonstrates robust performance across diverse scenarios. In the intricate dance of architecture and feature extraction, MobileNetV3 emerges as a guiding light in pushing the boundaries of precision and efficiency for detection of apple flowers for pollination drones.

2.6 *Modified Neck and Head of YOLOv5*

2.6.1 Improved Neck and Head: GhostNet Integration

The integration of GhostNet into YOLOv5s for apple flower detection goes beyond technical considerations; it's a harmonious design symphony [10, 31]. GhostNet's lightweight nature complements the efficiency of MobileNetV3, crafting a neural network finely attuned to the intricacies of apple flowers. The nuanced choreography between the backbone and the GhostNet-laden head ensures a graceful dance through the feature space, capturing the essence of apple flowers with precision. GhostNet's architecture plays a pivotal role in minimizing computational costs, addressing challenges posed by traditional deep CNNs. While methods like MobileNet and ShuffleNet introduced depthwise convolution and shuffle operations for efficiency, [10, 31] GhostNet innovatively focuses on reducing redundancy in intermediate feature maps. This approach enhances the model's efficiency and makes GhostNet an instrumental choice for the specialized task of apple flower detection within the YOLOv5s model.

2.6.2 GhostNet's Role in Computational Efficiency

Consider the operation of a convolutional layer for producing n, feature maps from input data Eq. 4, as shown in Fig. 17, where, c, is the number of input channels and h, and w, are the height and width respectively.

$$X \in \mathbb{R}^{c \times h \times w} \tag{4}$$

This operation can be expressed as Eq. 5, where, Eq. 6, are the convolution filters and, b is a bias term [10, 31]. The GhostNet strategy introduces intrinsic feature maps. Equation 7, generated through primary convolution Eq. 8. Equation 9, shows utilized filters, where, $m \leq n$. and m, is the number of intrinsic feature maps generated by the GhostNet strategy, and n is the total number of feature maps produced by the convolutional layer. The constraint m $\leq$ n ensures that the GhostNet strategy reduces the number of feature maps for computational efficiency.

$$Y = X * f + b \tag{5}$$

$$f \in \mathbb{R}^{c \times k \times k \times w} \tag{6}$$

$$Y' \in \mathbb{R}^{h' \times w' \times m} \tag{7}$$

$$Y' = X * f' \tag{8}$$

$$f' \in \mathbb{R}^{c \times k \times k \times m} \tag{9}$$

The convolution filters Eq. 10, play a pivotal role in the overall process [10, 31]. Specifically, c, denotes the number of input channels, $k \times k$ represents the spatial dimensions of each filter, and w signifies the number of filters. This configuration captures the spatial relationships within the input data, allowing the network to learn hierarchical features effectively. The reduced-depth filters Eq. 11, in the GhostNet strategy contribute to parameter reduction and computational efficiency, facilitating improved performance in detection pipelines.

$$f \in \mathbb{R}^{c \times k \times k \times w} \tag{10}$$

$$f' \in \mathbb{R}^{c \times k \times k \times m} \tag{11}$$

2.6.3 Linear Operations (Φ) for Final Feature Maps

To obtain n feature maps Eq. 12, GhostNet applies a series of cheap linear operations (Φ) on each intrinsic feature in Y as shown in Fig. 18.

$$Y = [y_{11}, y_{12}, \ldots, y_{ms}] \tag{12}$$

$$y_{ij} = \Phi_{i,j}(y_{i'})\forall i = 1, \ldots, m, \ j = 1, \ldots, s, \tag{13}$$

The last operation $\Phi_{i,s,}$ is the identity mapping, preserving the intrinsic feature maps. This approach significantly reduces the number of parameters and FLOPs, enhancing the efficiency of the overall detection pipeline.

2.6.4 In-depth Explanation of Head Module Layers

The head module plays a critical role in the detection of apple flowers by refining and integrating features from multiple scales. Let's delve into an in-depth explanation of each layer and its contribution to the overall goal of accurate apple flower detection:

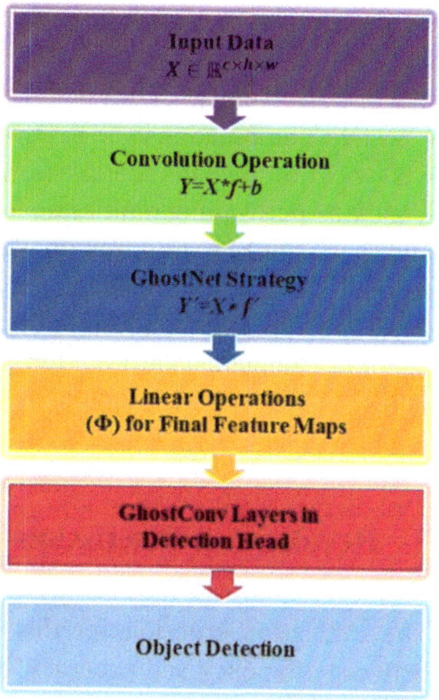

Fig. 18 Feature extraction process of GhostNet

Layer 12 introduces GhostConv, a 1 × 1 convolutional operation with 512 filters, aimed at reducing dimensionality and channel size to focus on essential information for object detection. Layer 13 employs nn. Upsample using nearest-neighbor interpolation to double spatial resolution, recovering lost spatial information. Layer 14 performs concatenation, combining upsampled features with backbone features at P4 level, enhancing object localization and capturing fine details in apple flowers. Layer 15 features C3Ghost, a composite convolutional block without batch normalization, applied to concatenated features. This enhances the network's ability to capture complex patterns, contributing to understanding intricate structures within apple flowers. Layers 16–19 replicate earlier operations on features from P3, further refining features and capturing both low and high-level characteristics. Layers 20–25 follow a similar pattern on higher-level features (P4 and P5), emphasizing larger context with GhostConv using a larger kernel and stride. Layer 26, the final layer, focuses on object detection. It utilizes features from multiple scales (P3, P4, P5) and anchor boxes to predict bounding boxes, abjectness' scores, and class probabilities. The network's learned associations enable precise detection and classification of apple flowers. In summary, these layers collectively form a comprehensive pipeline, leveraging GhostConv, upsampling, concatenation, and C3Ghost blocks to refine features, recover spatial information, and enable accurate detection of intricate details in apple flower images.

GhostNet, seamlessly integrated with YOLOv5's detection head alongside MobileNetV3, strategically refines features for efficient and precise apple flower detection. The GhostConv layers optimize channel usage, ensuring seamless feature integration across scales (P4 to P5). This design choice not only prioritizes computational efficiency but also acts as a neural refinement engine, enhancing the model's discernment of fine details crucial for accurate detection. The MobileNetV3-GhostNet combination in YOLOv5 is tailored for real-time object detection, aligning well with the demands of apple flower detection. The head module contributes significantly by performing multiscale feature fusion (P3, P4, P5), capturing both low and high-level semantic information with GhostConv and C3Ghost blocks. Spatial refinement is achieved through strategic upsampling and concatenation, emphasizing fine details in apple flowers. Additionally, GhostConv's larger kernel sizes and strides facilitate contextual information integration, enhancing the model's capability to detect apple flowers in complex scenes.

3 Results and Discussions

The evaluation criterion for this study emphasizes achieving a delicate balance between efficiency and accuracy. The focus is on ensuring a lightweight model (as indicated by number of parameters, GFLOPs, FPS and model size) while maintaining high detection efficiency. The goal is to meet the requirement of low complexity for embedded devices, making the model suitable for real-world deployment. To strike this balance, the evaluation prioritizes accuracy (as reflected by mAP, P, R, and F1 score) with a target of achieving an mAP50 score exceeding 90%. This meticulous selection of evaluation metrics aligns with the overarching objective of creating a model tailored for efficient apple flower detection in natural environments, with applicability to embedded devices like Jetson or Raspberry Pie.

3.1 YOLOv5 Models Comparison

3.1.1 Modified YOLOv5s with MobileNetV3 Backbone and GhostNet Neck

The modification of YOLOv5s with a MobileNetV3 backbone and GhostNet Neck represents a strategic enhancement aimed at optimizing the trade-off between efficiency and accuracy. The specific architecture changes led to noteworthy results in precision, recall, and Average Precision (AP), as shown in Table 2.

3.1.2 YOLOv5s Models Comparative Analysis

Precision, Recall and mean Average Precision (mAP50): The slightly reduced precision and recall are well-justified by the substantial reduction in GFLOPs and model size. The model maintains a commendable level of accuracy while significantly enhancing efficiency.

Efficiency Metrics: The drastic reduction in GFLOPs (from 77.4 to 8.5) and model size (from 67.0 to 9.7 MB) is a pivotal achievement from Table 4. These reductions signify a substantial improvement in computational efficiency and storage requirements, making the model apt for resource-constrained devices like Jetson and Raspberry Pie.

The percentage performance comparison illustrated in Tables 3 and 4 elucidates the superiority of the Modified YOLOv5 over its counterparts. This model achieves a comparable level of accuracy while significantly improving efficiency metrics. Notably, the reduction in GFLOPs and model size by approximately 89% and 85%, respectively, showcases the transformative impact of the proposed modifications.

In the benchmarking analysis, three variants of the YOLOv5 model, namely YOLOv5s-Benchmark, YOLOv5s-with transformers backbone, and improved YOLOv5s with MobileNetV3 Backbone and GhostNet Head, were assessed for their resource utilization and processing efficiency. Notably, YOLOv5s-Benchmark and YOLOv5s-Transformers exhibited similar patterns with allocated memory reaching 8.2×10^9 bytes and processing times exceeding 300 min, translating to prolonged computational durations as shown in Fig. 19. Furthermore, both models consumed 95% of the GPU memory which can be seen in Fig. 20, indicating a substantial strain on computational resources as shown in Fig. 20. In contrast, the Modified YOLOv5s variant demonstrated a more resource-efficient behavior, with allocated bytes totaling 4.2×10^9, a processing time of approximately 127 min which can be observed in Fig. 19, and a GPU memory allocation of 51% as shown in Fig. 20. This suggests

Table 3 Accuracy comparison of YOLOv5s models

Accuracy	YOLOv5s-Benchmark	YOLOv5s-Transformers	Modified YOLOv5
Precision (P)%	93.25	94.02	90.06
Recall (R)%	88.31	87.82	87.69
F1-score%	91	91	89.42
Average precision (AP)%	94.02	94.09	91.37

Table 4 Efficiency comparison of YOLOv5s models

Efficiency	YOLOv5s-benchmark	YOLOv5s-transformers	Modified YOLOv5
Number of parameters	33,221,002	32,843,862	4,661,611
GFLOPs	77.4	75.3	8.5
Model size (MB)	67	66.2	9.7
Detection speed (ms)	5.4	6.2	3.5
FPS	156	138	227

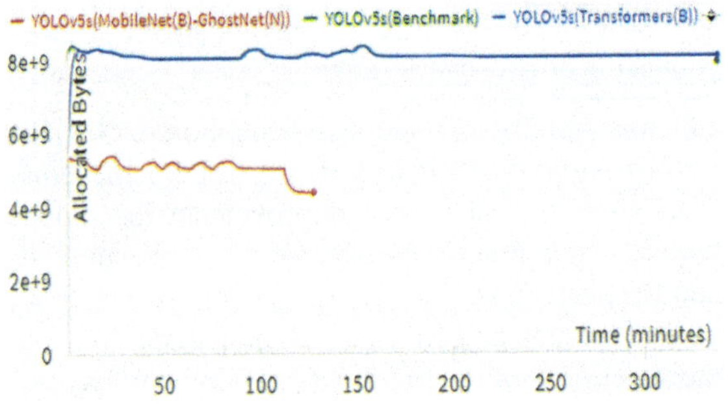

Fig. 19 System/GPU memory allocated Bytes

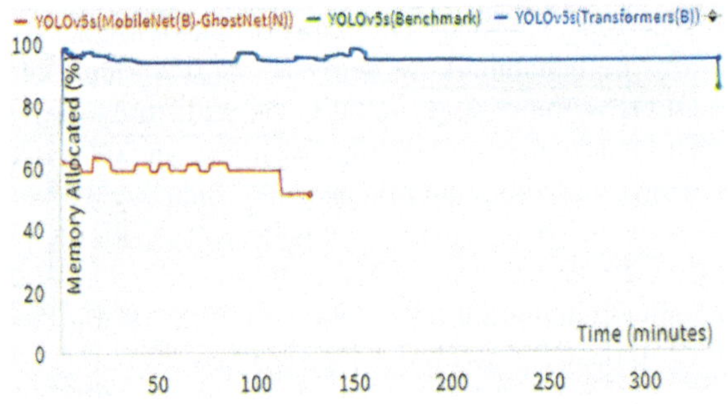

Fig. 20 Process GPU memory allocated (%)

that the modifications implemented in the latter model have led to a notable reduction in both processing time and GPU memory usage, potentially enhancing overall performance and computational efficiency. The findings underscore the significance of model optimization in achieving more streamlined and resource-conscious deep learning applications.

3.2 Comparative Analysis with State-of-the-Art Alternatives

3.2.1 YOLOv7

YOLOv7, with its precision of 99.1%, recall of 98.0%, and mAP50 of 99.0%, stands out as a high-accuracy model. However, it comes at the cost of a significantly higher computational load, with 103.2 GFLOPs and a model size of 74.8 MB as shown in Tables 5 and 6.

Table 5 Comparative analysis with state-of-the-art alternatives with modified YOLOv5

Accuracy	YOLOv7	Faster R-CNN	YOLOv8s	Modified YOLOv5
Precision (%)	99.1	–	87	90.06
Recall (%)	98	–	80	87.69
mAP50 (%)	99	79.9	89	91.37
F1-Score (%)	98.55	–	83.35	89.42

Table 6 Efficiency comparison of state-of-the-art alternatives with modified YOLOv5

Efficiency	YOLOv7	Faster R-CNN	YOLOv8s	Modified YOLOv5
GFLOPs	103.2	77.58	28.4	8.5
Model size (MB)	74.8	659.8	22.5	9.7
Number of parameters	36,481,772	–	11,125,971	4,661,611
FPS	120	51	220	227

3.2.2 YOLOv8s

YOLOv8s, lags in accuracy compared to YOLOv5s modified, and it is also not that efficient alternative with GFLOPs value of 28.4 and a model size of 22.5 MB as compared to YOLOv5s modified which can be observed in Tables 5 and 6.

3.2.3 Faster R-CNN

In Table 5 it can seen that faster R-CNN, with a precision of 44.0%, recall of 52.8%, and mAP50 of 79.9%, lags in accuracy compared to modified YOLOv5s. Its computational intensity is evident with a GFLOPs value of 77.58 and a large model size of 659.8 MB which is mention in Table 6.

3.2.4 Improved YOLOv5s

The Modified YOLOv5, with a precision of 90.06%, recall of 87.69%, and mAP50 of 91.37%, bridges the gap between YOLOv8s and YOLOv7 in terms of accuracy. Simultaneously, it excels in efficiency metrics with significantly lower GFLOPs (8.5) and model size (9.7 MB) faster FPS (190) and reduced number of parameters (4,661,611).

Fig. 21 Jetson Xavier NX

3.3 Execution Tests on Embedded Devices

The practical deployment of the improved YOLOv5s on Jetson Xavier NX as shown in Fig. 21 demonstrated a remarkable FPS rating of 19 FPS. This substantial improvement over the baseline YOLOv5s (10 FPS) is a testament to the model's efficacy in real-world, resource-constrained scenarios.

3.4 Discussion

The evaluation of the improved YOLOv5s incorporated with MobileNetV3 backbone and GhostNet head model reveals a meticulous balance achieved between efficiency and accuracy, aligning with the study's overarching goal of creating a lightweight model for efficient apple flower detection in natural environments. The focus on parameters such as Precision (P), Recall (R), Average Precision (AP), and F1 Score, combined with efficiency metrics like GFLOPs, model size, and detection speed, underscores the study's commitment to developing a model suitable for real-world deployment on embedded devices.

In the YOLOv5s Models Comparison, the modification with a MobileNetV3 backbone and GhostNet head results in a model that, despite a slight reduction in precision and recall, significantly improves efficiency. The drastic reduction in GFLOPs (from 77.4 to 8.5) and model size (from 67.0 to 9.7 MB) is a pivotal achievement, demonstrating the model's computational efficiency and reduced storage requirements.

The Comparative Analysis with State-of-the-Art Alternatives positions the Modified YOLOv5s as a compelling choice. It achieves a balance between accuracy and efficiency, outperforming alternatives like YOLOv7 and Faster R-CNN. Notably, it bridges the accuracy gap between YOLOv8s and YOLOv7 while excelling in efficiency metrics, with significantly lower GFLOPs and model size.

Execution tests conducted on embedded devices serve to further validate the practicality of the model. Impressively, the model achieves a remarkable Frames Per Second (FPS) rating of 19 FPS on the Jetson Xavier NX platform, underscoring its efficacy in resource-constrained scenarios. This is exemplified in Fig. 22, which portrays real-time multiple species of apple flowers detection results across diverse scenarios characterized by varying lighting conditions and flowering cluster complexities, ranging from relatively straightforward to intricate configurations.

Fig. 22 Apple flower detection results in multiple scenarios with improved YOLOv5s incorporated with MobileNetV3 backbone and GhostNet head **a**–**d** detection images in cloudy environment and with simple flower clusters. **e**–**h** detection images in shady environment and with relative complex flower clusters **i**–**l** detection images in bright sunny environment and with complex flower clusters

4 Conclusion

The meticulous modifications applied to YOLOv5 models, specifically with a MobileNetV3 backbone and GhostNet head, have yielded a model that stands out in the realm of efficient apple flower detection. The emphasis on striking a delicate balance between accuracy and efficiency has culminated in a solution that not only competes with but surpasses state-of-the-art alternatives. The Modified YOLOv5 exhibits commendable accuracy, with precision and recall values of 90.6% and 87.7%, respectively, and an impressive mAP50 of 91.2%.

Crucially, the model excels in efficiency metrics, achieving a transformative reduction in GFLOPs by approximately 89% and model size by 85%. This achievement is pivotal for real-world deployment on resource-constrained embedded devices, where computational efficiency and storage requirements are of utmost importance. The substantial improvement in detection speed, with the model achieving a remarkable 3.5 ms, further underscores its practicality in dynamic environments.

The Comparative Analysis with State-of-the-Art Alternatives positions the Modified YOLOv5 as a standout choice. It bridges the gap between models like YOLOv8s and YOLOv7, offering a compelling blend of accuracy and efficiency. The study also demonstrates the model's superiority over Faster R-CNN, known for its computational intensity and large model size.

Looking ahead, future directions could involve fine-tuning the model for specific environmental conditions or extending its applicability to diverse flora. Additionally, exploring edge computing optimizations and further reducing model size could enhance the model's deployment potential on even more resource-constrained devices.

In conclusion, the Modified YOLOv5 not only meets but exceeds the study's objective of achieving a delicate balance between accuracy and efficiency. This model presents a robust solution for real-world deployment, particularly in scenarios where lightweight, yet highly accurate, computer vision models are imperative. As technology evolves, further refinements and optimizations can undoubtedly enhance its effectiveness and broaden its applicability in various ecological and technological landscapes.

References

1. Zhu Z et al (2018) Life cycle assessment of conventional and organic apple production systems in China. J Clean Prod 201:156–168. https://doi.org/10.1016/j.jclepro.2018.08.032
2. Report T et al (2023) Report name: fresh deciduous fruit annual
3. Pardo A, Borges PAV (2020) Agriculture, ecosystems and environment worldwide importance of insect pollination in apple orchards: a review. Agric Ecosyst Environ 293(January):106839. https://doi.org/10.1016/j.agee.2020.106839
4. Losada JM, Herrero M (2013) Scientia horticulturae flower strategy and stigma performance in the apple inflorescence. Sci Hortic 150:283–289. https://doi.org/10.1016/j.scienta.2012.11.031

5. Dias PA, Tabb A, Medeiros H (2018) Computers in Industry Apple flower detection using deep convolutional networks. Comput Ind 99:17–28. https://doi.org/10.1016/j.compind.2018.03.010
6. Zhang C et al (2022) ScienceDirect automatic flower cluster estimation in apple orchards using aerial and ground based point clouds, vol 1. https://doi.org/10.1016/j.biosystemseng.2022.05.004
7. Bhattarai U, Karkee M (2022) A weakly-supervised approach for flower/fruit counting in apple orchards. Comput Ind 138:103635. https://doi.org/10.1016/j.compind.2022.103635
8. Wang XA, Tang J, Whitty M (2020) Side-view apple flower mapping using edge-based fully convolutional networks for variable rate chemical thinning. Comput Electron Agric 178(April):105673. https://doi.org/10.1016/j.compag.2020.105673
9. Wu D, Lv S, Jiang M, Song H (2020) Using channel pruning-based YOLO v4 deep learning algorithm for the real-time and accurate detection of apple flowers in natural environments. Comput Electron Agric 178(September):105742. https://doi.org/10.1016/j.compag.2020.105742
10. Shang Y, Xu X, Jiao Y, Wang Z, Hua Z, Song H (2022) Using lightweight deep learning algorithm for real-time detection of apple flowers in natural environments. Comput Electron Agric 207(September 2022):107765. https://doi.org/10.1016/j.compag.2023.107765
11. Tian Y, Yang G, Wang Z, Li E, Liang Z (2020) ScienceDirect instance segmentation of apple flowers using the improved mask R e CNN model, vol 3. https://doi.org/10.1016/j.biosystemseng.2020.03.008
12. Temple TN, Johnson KB, Pathology P (2010) Evaluation of loop-mediated isothermal amplification for rapid detection of *Erwinia amylovora* on pear and apple fruit flowers, no November 2010, pp 423–430
13. Mu X, He L, Heinemann P, Schupp J, Karkee M (2023) Smart agricultural technology mask R-CNN based apple flower detection and king flower identification for precision pollination. Smart Agric Technol 4(December 2022):100151. https://doi.org/10.1016/j.atech.2022.100151
14. Yuan W (2023) Accuracy comparison of YOLOv7 and YOLOv4 regarding image annotation quality for apple flower bud classification, pp 413–424
15. Chen Z et al (2022) Automatic estimation of apple orchard blooming levels using the improved YOLOv5
16. Zhang Y, He S, Wa S, Zong Z, Liu Y (2021) Using generative module and pruning inference for the fast and accurate detection of apple flower in natural environments, pp 1–16
17. Dias PA, Tabb A, Medeiros H (2018) Multispecies fruit flower detection using a refined semantic segmentation network. IEEE Robot Autom Lett 3(4):3003–3010. https://doi.org/10.1109/LRA.2018.2849498
18. Hočevar M, Širok B, Godeša T, Stopar M (2014) Flowering estimation in apple orchards by image analysis. Precis Agric 15(4):466–478. https://doi.org/10.1007/s11119-013-9341-6
19. Farjon G, Krikeb O, Hillel AB, Alchanatis V (2020) Detection and counting of flowers on apple trees for better chemical thinning decisions. Precis Agric 21(3):503–521. https://doi.org/10.1007/s11119-019-09679-1
20. Yuan W, Choi D, Bolkas D, Heinemann PH, He L (2022) Sensitivity examination of YOLOv4 regarding test image distortion and training dataset attribute for apple flower bud classification. Int J Remote Sens 43(8):3106–3130. https://doi.org/10.1080/01431161.2022.2085069
21. Aggelopoulou AD, Bochtis D, Fountas S, Swain KC, Gemtos TA, Nanos GD (2011) Yield prediction in apple orchards based on image processing. Precis Agric 12(3):448–456. https://doi.org/10.1007/s11119-010-9187-0
22. Zeng T, Li S, Song Q, Zhong F, Wei X (2023) Lightweight tomato real-time detection method based on improved YOLO and mobile deployment. Comput Electron Agric 205(November 2022):107625. https://doi.org/10.1016/j.compag.2023.107625
23. Guo G, Zhang Z (2022) Road damage detection algorithm for improved YOLOv5. Sci Rep 12(1):1–12. https://doi.org/10.1038/s41598-022-19674-8
24. Liu K, Wang J, Zhang K, Chen M, Zhao H, Liao J (2023) A lightweight recognition method for rice growth period based on improved YOLOv5s. Sensors 23(15). https://doi.org/10.3390/s23156738

25. Yu K, Tang G, Chen W, Hu S, Li Y, Gong H (2023) MobileNet-YOLO v5s: an improved lightweight method for real-time detection of sugarcane stem nodes in complex natural environments. IEEE Access 11(September):104070–104083. https://doi.org/10.1109/ACCESS.2023.3317951
26. Zhang L, Zhao C, Feng Y, Li D (2023)"Pests Identification of IP102 by YOLOv5 embedded with the novel lightweight module. Agronomy 13(6). https://doi.org/10.3390/agronomy13061583
27. Li S, Zhang S, Xue J, Sun H (2022) Lightweight target detection for the field flat jujube based on improved YOLOv5. Comput Electron Agric 202(April):107391. https://doi.org/10.1016/j.compag.2022.107391
28. Wen Z et al (2023) A lightweight small object detection algorithm based on improved YOLOv5 for driving scenarios. Int J Multimed Inf Retr 12(2):1–14. https://doi.org/10.1007/s13735-023-00305-5
29. Howard A et al (2019) Searching for mobileNetV3. Proc IEEE Int Conf Comput Vis 2019(Octob):1314–1324. https://doi.org/10.1109/ICCV.2019.00140
30. Liu S, Mu X, Chen Y, Zhang G (2023) Detection of Speckle defects in alumina ceramic balls based on MobileNetV3-YOLOv5. In: 2023 4th international conference on big data, artificial intelligence and internet of things engineering. ICBAIE 2023, pp 181–185. https://doi.org/10.1109/ICBAIE59714.2023.10281277
31. Han K, Wang Y, Tian Q, Guo J, Xu C, Xu C (2020) GhostNet: more features from cheap operations. Proc IEEE Comput Soc Conf Comput Vis Pattern Recogn 1577–1586. https://doi.org/10.1109/CVPR42600.2020.00165
32. Chollet F (2017) Xception: deep learning with depthwise separable convolutions. In: Proceedings, 30th IEEE conference on computer vision and pattern recognition, CVPR 2017, vol 2017-Janua, pp 1800–1807. https://doi.org/10.1109/CVPR.2017.195
33. Hu J, Shen L, Sun G (2018) Squeeze-and-excitation networks. Proc IEEE Comput Soc Conf Comput Vis Pattern Recogn 7132–7141. https://doi.org/10.1109/CVPR.2018.00745

Apple Size Estimation Method with 3D Projection Correction for In-field Grading System

Jiangfan Yu, Zhao Zhang, Yifei Li, Can Hu, and Xufeng Wang

Abstract Apples postharvest sorting is crucial for increasing economic benefits. This study designed an apple grading system for in-field sorting, consisting of a low-cost wide-field camera, bar-shaped light sources, screw conveyors, and a computer system. An automatic method for estimating the size of moving apples based on two primary steps was developed: (1) detecting the maximum cross-sectional diameter of the apple based on its orientation, and (2) using a 3D projection correction algorithm to adjust the detected diameter. First, each apple was tracked, and its stem/calyx region was detected to determine its orientation. Then, the maximum cross-sectional diameter of the apple in the 2D image was obtained based on orientation. Finally, 3D correction equation was constructed to calibrate the diameter, addressing the projection distortion caused by the apple being away from the camera. Sixty 'Fuji' and sixty 'Golden Delicious' apples were used for the experiments. The orientation detection accuracy of the algorithm was 93.1% for 'Fuji' apples and 98.5% for 'Golden Delicious' apples. For size estimation, the percentage of apples with an error within ± 5 mm was 98.3% ('Fuji') and 100% ('Golden Delicious'). Furthermore, the percentage of apples with an error within ± 2 mm was 75% ('Fuji') and 91.7% ('Golden Delicious'), with an overall RMSE of 1.62 mm. The experimental results indicate that the proposed method can accurately estimate the size of apples despite changes in their posture and position during movement. This performance surpasses that of similar studies and provides technical support for apple size grading. The

J. Yu · Z. Zhang (✉) · Y. Li
Key Laboratory of Smart Agriculture System Integration, Ministry of Education, Beijing 100083, China
e-mail: zhaozhangcau@cau.edu.cn

Key Laboratory of Agricultural Information Acquisition Technology, Ministry of Agriculture and Rural Affairs, Beijing 100083, China

College of Information and Electrical Engineering, China Agricultural University, Beijing 100083, China

C. Hu · X. Wang
College of Mechanical and Electronic Engineering, Tarim University, Alar 843300, Xinjiang, China

Y. Liu et al. (eds.), *New Technologies Applied in Apple Production*, Smart Agriculture 10, https://doi.org/10.1007/978-981-97-7778-5_5

method also shows potential for application in sorting other near-spherical fruits and vegetables.

Keywords Apples postharvest sorting · Size estimation · Orientation · Maximum cross-sectional diameter · Projection correction

1 Highlights

- Apple in-field grading system was developed
- An orientation based apple size estimation method was proposed
- 3D projection correction algorithm was constructed based on 2D wide field of view (FoV) image
- Apple size estimation error reduced to within ± 2 mm for in-field grading system

2 Introduction

Apples are one of the most produced fruits worldwide, ranking third in terms of output (FAO 2023). Due to variations in the external quality of apples, growers typically grade them before sale, which maximizes revenue based on the grading results [1, 5, 25]. Common grading criteria include size, color, shape, and defects, with size being the primary grading criterion [18, 23]. Current size grading methods include manual measurement and automated equipment detection [16]. In production, manual measurement methods suffer from visual variability, high subjectivity, low efficiency, and high costs, adversely affecting the economic efficiency of the apple industry [2, 10, 12, 15]. Therefore, developing automatic detection algorithms for apple size is of significant importance.

Researchers have proposed various machine vision based algorithms for automatic apple size detection. Most algorithms use geometric parameters of the apple to represent its size, such as fitting an ellipse to the contour and extracting its major axis (Throop 2003), extracting the length of the bounding rectangle [19], constructing a 3D model of the apple based on point clouds [22], extracting the diameter of the circumscribed circle [24], and extracting the maximum diameter of the contour [27]. However, these methods do not estimate apple size based on the maximum cross-sectional diameter as specified in standards ([21], GB/T 10651-2008). According to these standards, the maximum cross-sectional diameter of an apple is defined as the maximum diameter perpendicular to the axis connecting the stem and calyx. Due to the irregular shape of apples and the difficulty in detecting the stem and calyx regions, extracting the maximum cross-sectional diameter presents a significant challenge. For detecting the maximum cross-sectional diameter of apples, Chen et al. [3] extracted the maximum diameter of the apple and established a multiple linear regression model to fit the maximum cross-sectional diameter, achieving a grading accuracy of 87.1%. Ma et al. [14] used 3D reconstruction technology to find the slice with the largest cross-sectional area in the apple model and used the maximum diameter of this slice as the apple size, achieving an R2 of 0.95. This method performs well when the apple orientation is perpendicular to the slicing direction. However, in the actual grading process, apples are irregular and moving [7], and they do not remain in a fixed posture under the camera, leading to significant detection errors.

In addition to not using the maximum cross-sectional diameter as the apple size and not considering the movement of apples, previous studies also ignored the issue of projection distortion caused by changes in the distance between the apple and the camera center [9, 11]. The images obtained by machine vision are 2D projections of 3D objects. When an apple faces the camera's optical axis, its projected diameter is close to the actual value. However, under high throughput demands, the camera's field of view is typically large to cover more apples. During movement, apples deviate from the optical axis, causing projection distortion and resulting in a detected diameter that does not match the actual value. Mizushima and Lu [17] pre-used two spheres of different diameters moving on a conveyor to record the unit pixel size at different positions, fitting a function based on the statistical data for estimating the unit pixel size of apples at any position to compensate for projection distortion, with over 98.9% of apples having an error within $\pm$ 5 mm. However, this method uses two local special cases instead of the whole, not fundamentally solving the distortion problem. The diameters of the spheres are fixed, while the size range of apples is large, making the fitted function unsuitable for every apple. Additionally, this method performs better when the camera coverage is small, with increasing errors as the coverage area expands.

In summary, current machine vision-based apple size detection methods have two main issues: firstly, most methods do not use the maximum cross-sectional diameter as defined in national standards to determine apple size, failing to meet grading standards. Secondly, most methods do not consider the movement of apples, with their posture and position constantly changing during movement. Posture changes make detecting the maximum cross-sectional diameter difficult, and position changes lead to detection distortion.

This study proposes a machine vision-based method for detecting apple size, with specific objectives including: (1) assembling a low-cost in-field grading system for apple size estimation; (2) developing a method for detecting the maximum cross-sectional diameter based on apple orientation, considering the common post-harvest condition of apples having no or short stems; (3) developing an adaptive projection correction algorithm based on theoretical analysis to address projection distortion caused by changes in apple position, aiming to improve the accuracy of apple size estimation.

3 Materials and Methods

3.1 In-field Grading System

The apple field grading system (Fig. 1a) consists of an industrial camera (MV-CB013-A0UM-C, Hikrobot, Hangzhou, Zhejiang, China) and six bar-shaped light sources (LSN-BLI22045-W, Dongguan, Shenzhen, China). Each channel is formed by two screw conveyors, equipped with variable-pitch threads to ensure the separation and singulation of the apples [13]. In addition to transporting the apples, the two conveyors rotate in the same direction, using friction with the apples to rotate them, thereby ensuring that the grading system captures the complete surface information of the apples. The camera is fixed at a height of 537 mm from the conveyors, covering an area of 800 mm × 600 mm, allowing it to simultaneously capture motion videos of apples in three channels. The camera operates at a frame rate of 20 FPS, with apples moving within the field of view for 4–6 s, allowing each apple to be captured in at least 80 frames. To improve image quality and ensure segmentation accuracy, six bar-shaped light sources are installed on both sides of the apple movement direction for illumination. A CPU (Intel i5-13490F, Santa Clara, California, U.S.) and a GPU (RTX 4070, Santa Clara, California, U.S.) are used to process the captured videos, and the apple size estimation algorithm is developed using C++ and the OpenCV library. A total of 60 'Fuji' and 60 'Golden Delicious' apples were used in the experiments (Fig. 1b), with diameters ranging from 70 to 110 mm, covering the common sizes of apples in the market.

(a) (b)

Fig. 1 **a** In-field grading system and **b** 'Fuji' and 'Golden Delicious' for experiment

3.2 Image Preprocess

During the imaging process, imperfections in lens manufacturing can cause image distortion. Radial distortion is the most common type of lens distortion and has the greatest impact on image quality [4, 20]. The further an object is from the center of the camera, the more severe the distortion becomes, leading to errors in the final size detection of apples. Therefore, to ensure the accuracy of the detection system, the primary task before detecting and tracking apples is to correct the camera distortion.

The Zhang's calibration method [26] uses multiple chessboard images for camera calibration. The essence of this method is to establish a correspondence between the real-world object points and the image points on the 2D imaging plane. Compared to other calibration methods, Zhang's calibration method is simple to operate and highly accurate, suitable for most camera calibration scenarios. In this study, Zhang's calibration method was used to correct the camera distortion. A chessboard with a grid size of 9×7 and a square size of 24 mm was used for camera calibration. The camera was fixed in position, and 20 photos were taken by adjusting the position and angle of the chessboard multiple times. MATLAB (The Mathworks, Inc., Natick, Mass., USA) was used to calculate the camera's intrinsic and extrinsic matrix as well as the distortion coefficients. These parameters were then used to correct the subsequent video frames. After image correction, images of the chessboard at the edges and center of the camera's field of view were taken, resulting in three images: left, center, and right. The actual length L_{R} of the chessboard and the pixel length L_{I} in the images were used to calculate the unit pixel length P. The average of these measurements was taken as the final unit pixel length, which was determined to be 0.51 mm/pixel. The formula is given by:

$$P = \frac{L_{\mathrm{R}}}{L_{\mathrm{I}}} \tag{1}$$

where P is the unit pixel length in mm/pixel, L_{R} is the actual length of the chessboard in mm, and L_{I} is the pixel length of the chessboard in the image in pixels.

3.3 Apple Detection and Tracking

To achieve high throughput in-field grading of apples, it is crucial to quickly and accurately detect and track multiple objects. In the grading process, imaging systems typically minimize background interference as much as possible. In this study, the background of the imaging system was set to black, which has a high contrast with the color of apples. However, due to the smooth materials on some parts of the conveyors, specular reflection can occur when light shines on their surfaces, causing strong white light and color distortion, which leads to interference. The V channel of the HSV (hue, saturation, and value) color space is very sensitive to brightness and was thus selected for image processing. After preliminary color segmentation, morphological processing methods such as small area removal, opening operation, and hole filling were used to extract the apple regions.

After detecting each apple, it is essential to ensure that each apple in the video receives the same size label across different frames, requiring the apples in consecutive frames to be matched to achieve multiple object tracking (MOT). Models such as Kalman filters and particle filters handle the tracking problem through motion prediction by predicting the object's position in the next frame and correcting the result based on the object's features at that position. These models perform well for objects with regular motion patterns. However, apples on a spiral conveyor exhibit irregular motion due to the temporary thrust of the threads, causing acceleration, deceleration, and occasional stalling or retreating, leading to poor tracking performance with prediction-based methods. This can result in tracking drift or ID switching, reducing grading accuracy. Although data association algorithms added to Kalman filtering can address these issues, they increase algorithm complexity, making them unsuitable for low-cost, high-throughput field grading machines. Generative tracking algorithms such as meanshift and DFT extract the object features labeled in the first frame and match the most similar features in subsequent frames. These methods focus on the object's features, but their performance deteriorates when the object's features change significantly. Since apples undergo rotational motion, the features presented to the camera vary across frames, making these methods unsuitable for apple grading. Emerging convolutional neural network-based tracking methods can extract stable features of targets in various motion conditions, offering excellent tracking performance. However, these methods require extensive offline training, are inefficient, and have high costs.

Therefore, this study developed a fast, low-cost MOT algorithm. This algorithm combines simple color threshold segmentation with a maximum weight matching method based on Euclidean distance features to achieve MOT of apples while minimizing algorithm complexity and improving detection efficiency. After detecting apples in each frame, the centroids of the apples are extracted, and maximum weight matching is used to associate centroids across frames, assigning the same ID to the same apple. Unmatched apples, including new entries into the field of view and those that disappear, are distinguished by checking if an ID was assigned to them in the previous frame.

3.4 Apple Size Estimation

The camera captures video at a frame rate of 20 FPS. While tracking the apples, an image is extracted every 5 frames to estimate the size of the apples. During the movement of each apple, 16 images can be extracted. These images are used to detect the apple's S/C region and determine the apple's orientation. Based on the orientation, the maximum cross-sectional diameter is determined, and finally, the projection correction algorithm is applied to obtain the apple size. If the apple's S/C region is not detected, the estimation process is skipped.

3.4.1 Apple Orientation

The orientation of an apple aligns with the axis connecting its stem and calyx. However, in images, it is challenging to simultaneously capture both the stem and calyx regions, making it difficult to directly locate these areas. Given the nearly spherical nature of apples, the stem/calyx (S/C) axis approximately passes through the center of the sphere, exhibiting symmetrical properties that allow the apple to be divided into two relatively symmetrical halves. Therefore, the orientation of the apple can be determined by exploiting this symmetry. Mizushima and Lu [17] divided the apple contour into two segments and calculated the autocorrelation of the curve functions of these segments. The line connecting the endpoints of the pair of curves with the highest autocorrelation is considered the apple's orientation. However, due to the irregular shape of apples, the symmetry axis found using this method does not necessarily pass through the S/C regions, directly affecting the calculation of the maximum cross-sectional diameter. Therefore, when determining the apple's orientation, it is necessary to consider whether the selected symmetry axis passes through the center of the S/C region. The method proposed in this study involves two steps: (1) determining the center of the S/C region; (2) using symmetry to solve for the apple's orientation.

The stem and calyx regions are characterized by depressions, where the pixel intensity in images is usually lower, with the center of the depression typically having the lowest intensity. Methods such as color segmentation and edge detection are susceptible to interference from surface blemishes and defects, leading to misidentification of the S/C region. Further analysis of the pixel value distribution on the apple's surface revealed that the difference between the S/C regions and blemishes/defects lies in the fact that the R and G values of the S/C regions gradually decrease from the edge towards the center. Since the apple surface is red and yellow, the R channel is selected for further processing to enhance contrast. Visualization of the R channel shows that most of the surface area has $R > 200$, with the R value decreasing at the edges due to lower lighting (Fig. 2a). Additionally, there is a gradient area from red to green on the surface, with the R value corresponding to a gradient from 250 to 50, where the green area with the lowest R value represents the center of the S/C region. Based on this characteristic, determining the center of the S/C region can be transformed into a problem of finding the minimum value on a 2D surface. This problem can be solved by finding the minima of multiple arrays, obtained by traversing the 2D image row-wise and column-wise. For each array, the valley coordinates are determined (Fig. 2b). Finally, the average of the valley coordinates of all arrays is calculated as the center of the S/C region, with the calculated coordinates shown in Fig. 2c.

Once the center is determined, the next step is to find an appropriate symmetry axis as the apple's orientation. For a symmetrical object, the area on either side of

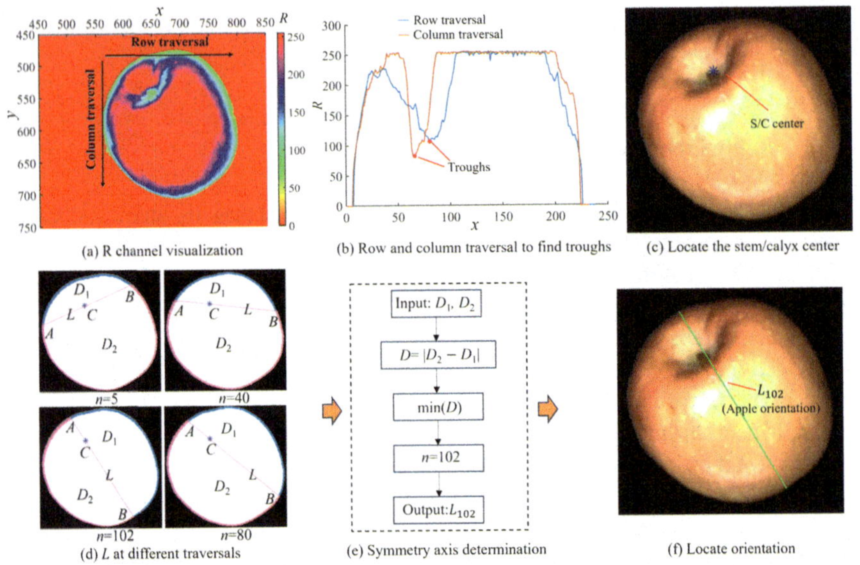

Fig. 2 Process of apple orientation determination. *A* and *C* are the starting point of traversal and S/C center; *L* is the line crossing *A* and *C*; *B* is the intersection of *L* and apple contour; D_1 and D_2, the area of the two-part of the apple divided by *L*; *D*, D_1 and D_2 minimum area difference; *n*, number of traversals

its symmetry axis should be equal. Thus, the region area can serve as an indicator of whether a line is a symmetry axis. The specific method to determine the symmetry axis is as follows:

(1) First, extract all the coordinate points of the apple's contour. Determine the starting point (point A) for contour traversal based on the value of n. Draw a line L through point A and the center of the stem/calyx region (point C), with the equation:

$$L: \frac{y - y_C}{y_A - y_C} = \frac{x - x_C}{x_A - x_C} \tag{2}$$

where x_A, y_A are the coordinates of point *A*, and x_C, y_C are the coordinates of point *C*.

(2) The line *L* will intersect the apple's contour at another point (point *B*). The line segment *AB* divides the apple into two parts: D_1 and D_2 (Fig. 2d). Calculate the area of these two parts and find the difference, denoted as *D*.

(3) Repeat the above steps for each contour point, resulting in multiple sets of points *A* and *B*, lines *L*, and differences *D*. After traversing all contour points, identify the *L* with the smallest difference *D*, which will be the apple's orientation (Fig. 2e).

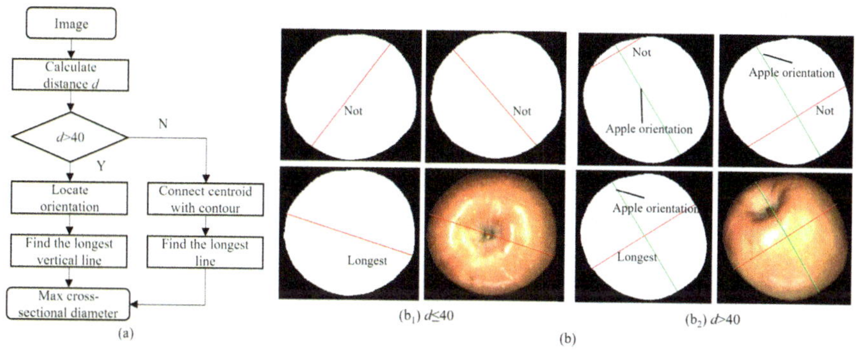

Fig. 3 **a** Detection process and **b** two diameter detection cases

In this example, when $n = 102$, meaning traversal has reached the 102nd contour point, the calculated line L best divides the apple into two parts, with the resulting apple orientation shown in Fig. 2f.

3.4.2 Maximum Cross-Sectional Diameter Estimation

The detection of the maximum cross-sectional diameter of an apple is based on the position of the S/C region and can be divided into two scenarios (Fig. 3a, b): If the Euclidean distance (d) between the S/C center and the apple's centroid is $\leq$ 40 pixels, it is assumed that the S/C axis is parallel to the camera's optical axis, indicating that the apple is in the upright position. In this case, the maximum cross-sectional diameter is the largest diameter within the apple's region, determined by the longest line connecting the centroid to the contour points of the apple. If the d is > 40 pixels, the apple is not in the upright position. Using the established apple orientation, the lines perpendicular to the apple's orientation are drawn from point A to point B. The longest line represents the maximum cross-sectional diameter. Due to projection distortion, the detected maximum cross-sectional diameter is not the true size of the apple. Therefore, the detected value must be corrected to obtain an accurate measurement.

3.4.3 Projection Correction

The images captured by the camera are 2D projections of 3D objects. Therefore, the maximum cross-sectional diameter detected in the image is not the true value. Figure 4 illustrates the projection scenario. When the apple is directly below the camera, the projection angle (φ) is small, making the projected diameter in the image close to the true value. Conversely, when the apple is far from the camera center, the projection angle (φ) is large, causing the projected diameter in the image to be larger than the true value. During the grading process, the position of the

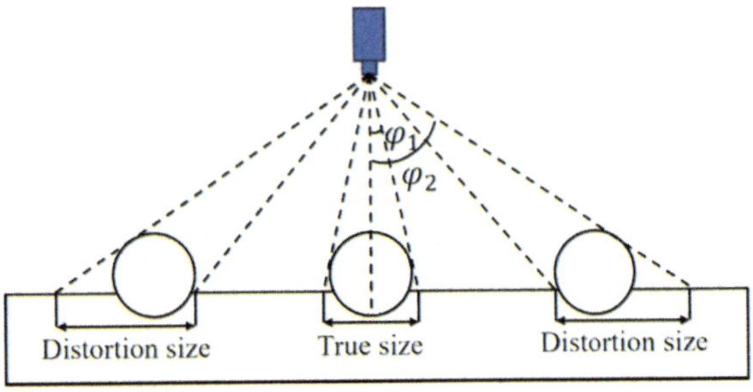

Fig. 4 Projection distortion principle

apple continuously changes, resulting in several images with large projection angles, which affects the detection results. To address this issue, we establish a relationship between the detected maximum cross-sectional diameter and the projection angle, constructing a projection equation to correct the detected diameter value.

To facilitate the analysis, this study uses a sphere to simulate the imaging mode of an apple. As shown in the Fig. 5a, at a certain moment, the sphere is located at a specific position in the camera's field of view. Figure 5b and c are the 2D image collected by camera and the 2D geometric diagram, respectively. Plane1 and Plane2 are constructed auxiliary planes. The horizontal plane where the top of the conveyors is located chosen as Plane1, serving as the imaging plane where the detected maximum cross-sectional diameter AB is situated. The plane where the center of the conveyor is located is selected as Plane2, due to the easily obtainable parameters like the distance between the conveyors and the diameter of the conveyors. It is important to note that Plane1 and Plane2 are adjustable, since their role is to provide geometric relationships related to the maximum cross-sectional diameter, with changes to these planes not affecting the final calculation results. Figure 5d shows the overall 3D projection diagram.

Based on Fig. 5d, an equation regarding the radius r can be obtained:

$$r = O\prime K \times \cos\varphi \tag{3}$$

where r is the radius of the sphere to be determined, O′K and φ are auxiliary variables.

Therefore, to obtain the true diameter of the sphere, it is necessary to determine O′K and φ. The subsequent derivation demonstrates how these two auxiliary variables can be obtained with known quantities using geometric relationships.

In Fig. 5d, it's easy to get:

$$\beta = \arctan\frac{HA}{MH} \tag{4}$$

$$\varphi = \arctan\frac{HB}{MH} \tag{5}$$

where β and φ are the left and right projection angle, °.

$$DB = MH \times (\tan\varphi - \tan\alpha) \tag{8}$$

From the right triangle MOH, HGG' and $O'CC'$, we can get MN:

$$MN = MH + \frac{MH}{MO}(HG - O'C) \tag{9}$$

where MO is the distance from the center of the camera to Plane1, HG is the distance from Plane1 to Plane2, and $O'C$ is the distance from the center of the sphere to Plane2.

Based on Fig. 5c, the $O'C$ can be obtained by an auxiliary equation:

$$O'C = \sqrt{(r+R)^2 - \left(R + \frac{D}{2}\right)^2} \tag{10}$$

Combining Eqs. (3–10), we can get the final correction equation about the radius r:

$$r = \left[MH + \frac{MH}{MO}\left(HG - \sqrt{(r+R)^2 - (R + \frac{D}{2})^2}\right)\right] \times \left[\frac{HB}{MH} - \tan\left(\frac{\arctan\frac{HA}{MH} + \arctan\frac{HB}{MH}}{2}\right)\right] \times \cos\left(\arctan\frac{HB}{MH}\right) \tag{11}$$

where MH can be obtained from MO and OH, OH is the distance from O to AB, HB and HA are the distances from H to the two endpoints B and A of the maximum cross-sectional diameter, R is the radius of the conveyor, and D is the two conveyors distance between. These are all known quantities, so r can be found, and the maximum cross-sectional diameter is twice r. Since the mechanical structure of each grading system is different, the auxiliary equation (Eq. 10) regarding r is different due to the different auxiliary planes selected, and the correction equations used cannot be directly applied to other grading systems. However, it is only necessary to change the auxiliary equations for different systems, and combine the other formulas derived above to obtain an update correction equation, which can be applied to any hierarchical system composed of mechanical structures, or even three-dimensional near-spherical systems in other fields.

4 Result and Discussion

The developed method first matches the centroids of multiple apples in the current frame with those in the previous frame based on maximum weight matching using Euclidean distance, enabling MOT. Then, it determines whether the S/C region can

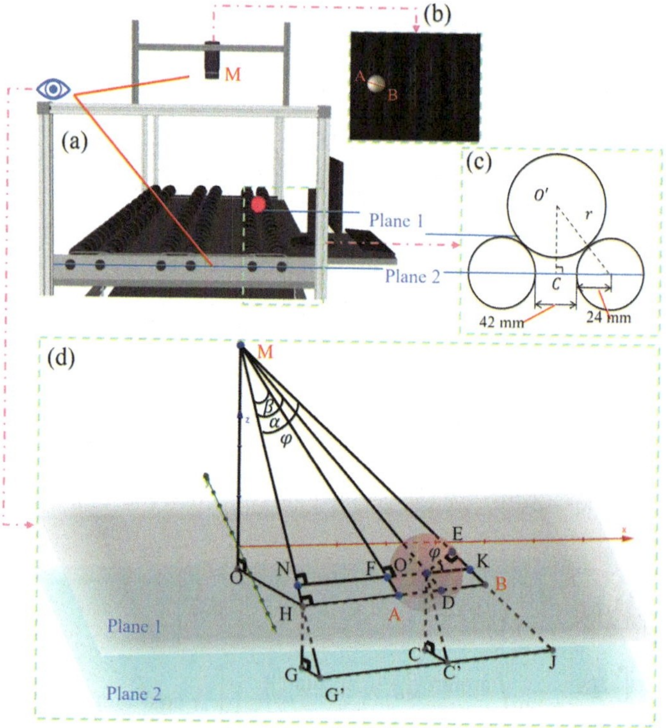

Fig. 5 **a** An example of imaging, **b** image by collecting, **c** front view from example and **d** 3d model for projection correction. E and F are the tangent points of MB and MA with the sphere, respectively, with $O'E$ being the radius r to be determined. N is the foot of the perpendicular from O' to MH, and point K is the intersection of the extension line of NO' and MB. For Plane 1, point O is the vertical projection of the camera's optical center M onto the plane, AB is the detected maximum cross-sectional diameter, point H is the foot of the perpendicular from O to line AB, and point D is the intersection of the extension line of MO' and the plane. For Plane 2, points G' and J are the intersections of the extension lines of MH and MB with the plane, respectively, while points G and C are the vertical projections of H and O' onto the plane. Point C' is the foot of the perpendicular from C to line G'J. From the above, it is evident that except for O, G, and C, all other points lie on plane $MG'J$

By applying the theory of similarity in triangles, we can get α and OK':

$$\alpha = \frac{\beta + \varphi}{2} \tag{6}$$

$$O'K = \frac{MN}{MH} \times DB \tag{7}$$

where α, MH, MH, DB are auxiliary variables. At this point, OK' can be obtained by determining MN, MH and DB.

From the right triangle MHD and MHB, we can get DB:

be detected. If the S/C region is located, the apple's orientation is determined based on symmetry, and size estimation with projection correction is performed; otherwise, the size estimation process is skipped. When the conveyor belt moves at a speed of 2 apples per second, no ID switching occurs. The image frame size captured by the video is 1440 × 1080. When there are 6 apples in the camera's field of view, and none of the S/C regions are detected (i.e., only MOT is performed), the processing speed can reach up to 80 fps. When size estimation and MOT are performed simultaneously for 6 apples, the processing speed is greater than 40 fps. In actual grading processes, reducing the image frame size can further increase the processing speed. Therefore, the developed algorithm fully meets the high-throughput, real-time requirements of practical field grading.

4.1 Apple Orientation Results

The axial direction of the stem and calyx connection represents the orientation of the apple. Table 1 presents the orientation estimation results for two types of apples, depicting the percentages of estimates within ± 10° and ± 20° error ranges using a symmetry-based orientation algorithm. Ground truth values were determined manually through visual observation.

In the ± 20° and ± 10° error ranges, the accuracy rates for 'Fuji' and 'Golden Delicious' apples were 93.1%, 84.4% and 98.5%, 87.8%, respectively. 'Golden Delicious' exhibited higher accuracy compared to 'Fuji' for two primary reasons: first, 'Golden Delicious' apples have a bright skin color, which enhances the differentiation from the stem and calyx region, thereby improving the accuracy of detection; second, 'Golden Delicious' apples have a rounder shape and lower deformity rate compared to 'Fuji', resulting in better performance when using symmetry to determine orientation. Errors beyond the ± 20° error range in orientation estimation are mostly attributed to cumulative errors in the localization of the stem and calyx region centers and irregularities in apple shape, leading to errors in symmetry axis determination.

Table 1 Results of orientation estimation

Apple variety	Orientation results			
	Total images	± 10° error (%)	± 20° error (%)	Out of ± 20° error (%)
Fuji	653	84.4	93.1	6.9
Golden Delicious	581	87.8	98.5	1.5

4.2 Projection Correction Validity

The projection correction equation was derived from analysis using ideal spheres. Therefore, experiments were conducted using two solid spheres of 79 and 89 mm sizes to validate the equation's effectiveness, representing the typical size range of most apples available on the market. Under constant camera height, the degree of projection distortion varies with the horizontal distance from the object to the camera optical axis. The experimental setup divided the camera capture area into nine regions labeled as upper, middle, and lower regions for each of the three channels. The spheres were sequentially placed in these nine sub-regions for image capture. Images obtained were processed with and without projection correction, and the results were compared as shown in the Fig. 6.

In Fig. 6b, orange indicates the corrected results, while green represents the results without projection correction. It is evident that the diameter detection errors for the spheres without projection correction are significant, with average relative errors of 19.4% and 21.5% for the 79 mm and 89 mm spheres, respectively, and standard deviations of 6.9 mm and 7.6 mm. After correction, the average relative errors decreased to 0.4 and 0.3%, with standard deviations of 1.0 and 1.2 mm. This validates the effectiveness of the projection correction equation. Additionally, the figure shows

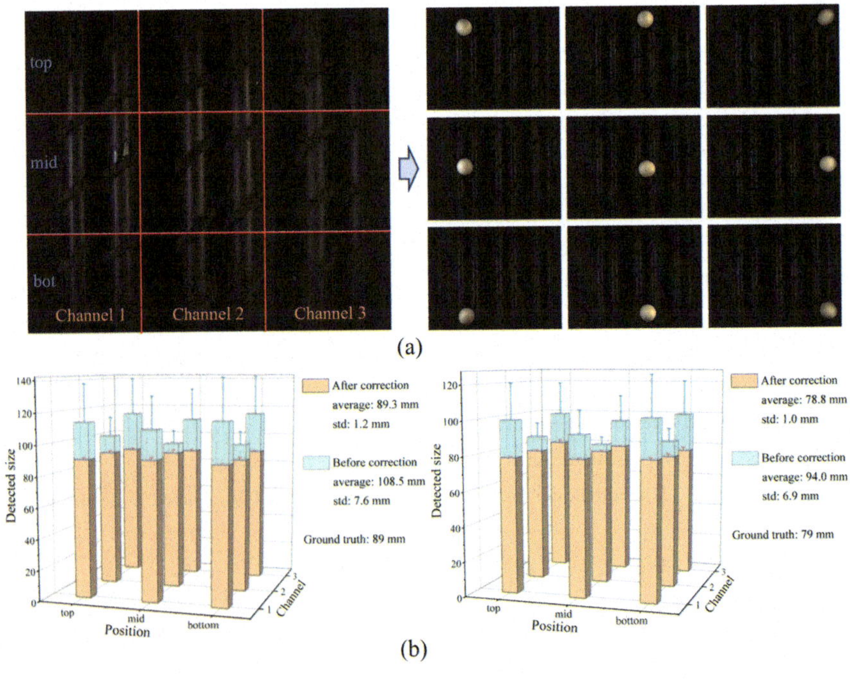

Fig. 6 **a** Nine sub-regions division scheme and **b** comparison of ball size detection in before and after correction

that the error in channel 2 before correction is relatively smaller compared to the other two channels, with the minimum error occurring in the middle. This is because the sphere's horizontal distance to the camera optical axis is the smallest in the middle position, resulting in less distortion. After correction, errors across all positions are reduced to a similar level, demonstrating the excellent performance of the projection correction equation.

4.3 Apple Size Results

The ground truth of apple size were measured manually using a vernier caliper (), oriented perpendicular to the axis formed by the apple's stem and calyx. Measurements were taken every 36 degrees around this axis, with the maximum recorded value selected as the true size of the apple. Table 2 presents the results of detecting the maximum cross-sectional diameter for 'Fuji' and 'Golden Delicious' apples. The Root Mean Square Error (RMSE) for both types of apples fall within an acceptable error range, with 'Golden Delicious' having an RMSE close to 1 mm.

The proposed method's detection errors are mostly within ± 5 mm, and to further evaluate its performance, the percentage of apples within ± 2 mm error is also listed. 'Golden Delicious' achieves a statistical result of 91.7%, while 'Fuji' achieves 75%. This variation is related to the shape and surface characteristics of the apples: Golden Delicious apples are more round and regular in shape, with a bright yellow surface that contrasts well with the background, facilitating more precise segmentation. Additionally, the separation between the stem and calyx region and the skin tissue is clearer, leading to more accurate orientation estimation, thereby affecting the accuracy of the maximum cross-sectional diameter estimation. On the other hand, 'Fuji' apples are irregular in shape and have a dark red surface, leading to inaccurate orientation estimation and deviations from the true values in the detection results. The Fig. 7 shows that most of Fuji's detection results are lower than the ground truth, primarily

Table 2 Results of maximum cross-sectional diameter estimation and comparison with other method

	RMSE (mm)	Within ± 2 mm error (%)	Within ± 5 mm error (%)
Method proposed			
Fuji	2.06	75.0	98.3
Golden delicious	1.17	91.7	100.0
Overall	1.62	83.4	99.2
Other method			
Mizushima and Lu [17]	1.79	–	98.9
Chen et al. [3]	–	–	87.1
Ma et al. [14]	1.74	–	–

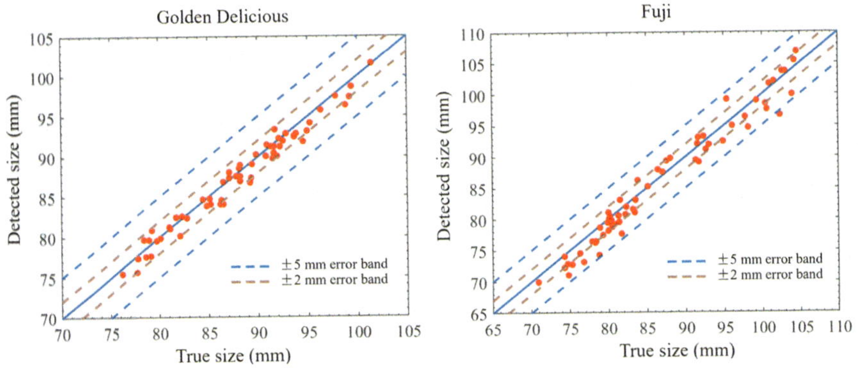

Fig. 7 Size estimation results of 'Golden Delicious' and 'Fuji'

due to edge pixels being misclassified as background, resulting in underestimation for most apples. In contrast, the detection results for Golden Delicious are distributed more evenly around the true values.

Compared to other similar studies on detecting the maximum cross-sectional diameter, the proposed method demonstrates satisfactory performance. It not only detects the sizes of apples in different positions and orientations during motion but also achieves lower RMSE values and higher detection accuracy compared to previous research. Moreover, the method further reduces errors to within ± 2 mm, a level of precision not reported in previous studies.

5 Conclusions

This study focuses on size estimation of apples in-field grading, addressing the challenges of varying apple posture and positions under a wide-field camera, as well as issues with projection distortion. A novel method based on apple orientation for detecting maximum cross-sectional diameters was developed, alongside a projection correction algorithm to mitigate diameter distortion caused by increasing projection angles. Performance experiments demonstrated that the algorithm achieves processing speeds up to 40 FPS for images containing six apples, meeting the sorting speed requirements for in-field. The orientation algorithm achieved accuracy rates of 93.1% for 'Fuji' and 98.5% for 'Golden Delicious' within a ± 20° error range. Superior estimation for 'Golden Delicious' was attributed to precise localization of stem and calyx positions and a more regular apple shape, facilitating accurate identification of the apple contour's symmetry axis. The size estimation algorithm estimated the maximum cross-sectional diameters for 'Fuji' and 'Golden Delicious' with an overall RMSE of 1.62 mm and an estimation success rate exceeding 99%. Furthermore, 75% ('Fuji') and 91.7% ('Golden Delicious') of apples were detected within ± 2 mm error, demonstrating satisfactory precision.

The proposed method addresses challenges in real-world grading scenarios where varying apple poses and positions affect detection accuracy, and projection angle distortion impacts measurement fidelity. It is adaptable for similar applications and related research involving nearly spherical fruits and vegetables, providing effective insights for grading. However, limitations include varying pixel lengths at different apple positions, necessitating further investigation into non-linear relationships between position and pixel length to enhance detection accuracy. Improvement is also needed in apple stem and calyx detection and contour segmentation, considering factors like lighting conditions and machine learning to bolster algorithm robustness. In addition, future enhancements should focus on throughput improvements for in-field grading systems.

References

1. Abbaspour-Gilandeh Y, Aghabara A, Davari M, Maja JM (2022) Feasibility of using computer vision and artificial intelligence techniques in detection of some apple pests and diseases. Appl Sci 12(2):906
2. Bhargava A, Bansal A (2021) Fruits and vegetables quality evaluation using computer vision: a review. J King Saud Univ Comput Inf Sci 33(3):243–257
3. Chen YJ, Zhang JX, Li W, Ren YX, Tan YZ (2012) Grading method of apple by maximum cross-sectional diameter based on computer vision. Trans Chin Soc Agric Eng 28(2):284–288
4. Duane CB (1971) Close-range camera calibration. Photogramm Eng 37(8):855–866
5. Fan SX, Liang XT, Huang WQ, Zhang VJL, Pang Q, He X, Li LJ, Zhang C (2022) Real-time defects detection for apple sorting using NIR cameras with pruning-based YOLOV4 network. Comput Electron Agric 193:106715
6. FAO. [FAOSTAT]. License: CC BY-NC-SA 3.0 IGO. Extracted from: https://www.fao.org/faostat/en/#data/QCL. Accessed 08 Jan 2024
7. Feng B, Wang MH (2003) Detecting method of fruit size based on computer vision. Trans Chin Soc Agric Mach 34(1):73–75
8. GB/T 10651—2008 (2008) Fresh apple. China Standard Press, Beijing
9. He DJ, Yang Q, Xue SP, Zhao J (1997) Computer vision method for measuring defect area of sphere-fruits. Trans Chin Soc Agric Eng 13(4):161–165
10. Hu X, Hu Y, Cai W, Xu Z, Zhao P, Liu X et al (2023) Automatic detection of small sample apple surface defects using ASDINet. Foods 12(6):1352
11. Liu T, Rui CJ, Li HT, Zhang SY (2019) Image distortion control and experiment of spherical fruit surface plaque. Trans Chin Soc Agric Mach 50(9):221–228
12. Lu R, Zhang Z, Pothula A (2017) Innovative technology for enhancing apple harvest and postharvest handling efficiency. Fruit Q 25(2):11–14
13. Lu Y, Lu R, Zhang Z (2022) Development and preliminary evaluation of a new apple harvest assist and in-field sorting machine. Appl Eng Agric 38(1):23–35
14. Ma H, Zhu X, Ji J, Wang H, Jin X, Zhao K (2021) Rapid estimation of apple phenotypic parameters based on 3D reconstruction. Int J Agric Biol Eng 14(5):180–188
15. Ma L, Wei P, Qu X, Bi S, Zhou Y, Shen T (2022) Apple grading method based on neural network with ordered partitions and evidential ensemble learning. CAAI Trans Intell Technol 7(4):561–569
16. Mhamed M, Zhang Z, Yu J, Li Y, Zhang M (2024) Advances in apple's automated orchard equipment: a comprehensive research. Comput Electron Agric 221:108926
17. Mizushima A, Lu R (2013) A low-cost color vision system for automatic estimation of apple fruit orientation and maximum equatorial diameter. Trans ASABE 56(3):813–827

18. Pothula AK, Zhang Z, Lu R (2023) Evaluation of a new apple in-field sorting system for fruit singulation, rotation and imaging. Comput Electron Agric 208:107789
19. Sofu MM, Er O, Kayacan MC, Cetişli B (2016) Design of an automatic apple sorting system using machine vision. Comput Electron Agric 127:395–405
20. Tsai R (1987) A versatile camera calibration technique for high-accuracy 3D machine vision metrology using off-the-shelf TV cameras and lenses. IEEE J Rob Autom 3(4):323–344
21. USDA (2019) United States Standards for Grades of Apple. USDA, Washington, DC
22. Wang HY, Yan RQ, Zhou XL, Ma SH, Hu HX, Xu HL (2019) Apple shape index estimation method based on local point cloud. Trans Chin Soc Agric Mach 50(5):205–213
23. Yang Y, Velastin SA, Fei Y (2020) Automatic grading of apples based on multi-features and weighted K-means clustering algorithm. Inf Process Agric 7(4):556–565
24. Zhang EY (2020) Research on key technologies of apple field grading. Northwest A & F University, Yangling
25. Zhang Z, Lu Y, Lu R (2021) Development and evaluation of an apple infield grading and sorting system. Postharvest Biol Technol 180:111588
26. Zhang Z (1999) Flexible camera calibration by viewing a plane from unknown orientations. In: Proceedings of the seventh IEEE international conference on computer vision. IEEE, Kerkyra, pp 666–673
27. Zhao DD, Ai Y (2022) Research on apple size detection method based on computer vision. J Agric Mech Res 44(7):206–209

Research on the Fruit Movement Mechanism of Apple In-field Sorting Equipment

Yifei Li, Jiangfan Yu, Zhao Zhang, N. S. Mustafa, and Dina Saber Salama

Abstract The apple in-field sorting system can reduce post-production costs and enhance the competitiveness of China's apple industry. Our laboratory has developed computerized imaging-based apple in-field sorting equipment consisting of a variable-pitch screw conveyor, a computerized imaging system, and a sorting actuator. To achieve efficient sorting of apples, the screw conveyor needs to simultaneously convey, separate, and uniformly rotate spherical-like fruits, ensuring that the imaging system quickly detects the full surface information of the apples. In this study, three sizes of apples, each coated with six different colors, were conveyed while the conveyor bar operated at 2 and 3 r/s. The variable-pitch conveyor was evaluated by developing a color-based surface segmentation algorithm for moving apples, calculating the average cumulative percentage, and investigating the number of rotational revolutions, the rotational motion pattern, and the horizontal motion distance of the apples on the variable-pitch conveyor rod at the two conveyor speeds. This clarified the relationship between apple size, conveyor speed, and horizontal motion distance. The results showed that the imaging system could image the surface of each apple more than 2–3 times, achieving an efficiency that allows processing and grading the surface information of 12 apples simultaneously within 1.0 s. As the pitch increases, the distance traveled by an apple during one complete rotation gradually increases. The larger pitch ensures that the full surface information of the apples is obtained,

Y. Li · J. Yu · Z. Zhang (✉)
Key Laboratory of Smart Agriculture System Integration of Ministry of Education, China Agricultural University, Beijing 100083, China
e-mail: zhaozhangcau@cau.edu.cn

Key Laboratory of Agricultural Information Acquisition Technology, Ministry of Agriculture and Rural Affairs, China Agricultural University, Beijing 100083, China

N. S. Mustafa
Pomology Department, Agricultural and Biological Research Institute, National Research Centre, Giza, Egypt

D. S. Salama
Water Relation and Field Irrigation Department, Agricultural and Biological Research Institute, National Research Centre, Giza, Egypt

Y. Liu et al. (eds.), *New Technologies Applied in Apple Production*, Smart Agriculture 10, https://doi.org/10.1007/978-981-97-7778-5_6

making the horizontal movement of the apples more uniform and smooth after separation. This leads to a more efficient and reliable acquisition of image information, providing a theoretical basis for realizing efficient in-field sorting of apples.

Keywords In-field sorting · Screw conveyor · Image segmentation · Average cumulative percentage · Apple

1 Introduction

Early post-harvest apples were mainly graded by manual grading to grade the quality of apples, but manual grading is not only low in productivity and high in cost, but its grading results are also susceptible to the influence of human subjective factors, and the accuracy fluctuates greatly, making it difficult to realize the highly demanding grading standards. With the help of advanced computer vision systems and fine artificial intelligence algorithms, the progress of fruit grading technology has led to a revolution in agricultural production, especially in the processing of apples and other fruits. Fruit sorting equipment from companies such as TOMRA Sorting Solutions in Norway, Aweta in the Netherlands, Compac in New Zealand, and Greenmeng in China are already being used in a wide range of commercial applications. These efficient fruit sorting machines are able to grade according to a variety of criteria such as weight, size, color, as well as internal quality (e.g., sugar and acidity) and surface defects, which greatly improves the accuracy and efficiency of sorting [1].

In order to effectively image and grade apples, the separation and rotation mechanism needs to simultaneously convey, separate, and uniformly rotate the quasi-spherical fruits to ensure that the conveyor mechanism meets the requirements of the imaging system for rapid detection of the complete surface information of the apples [2]. Separation and rotation of apples is a fundamental step in a computerized imaging-based grading system [3].

Fruit conveyors for fruit sorting systems have been widely developed over the years. Suzuki Motofumi et al. [4] utilized a reflective device to transfer fruits, but the device was complex in structure and could only cover at most 83% of the fruit surface, with a high rate of fruit surface leakage, which affected the grading accuracy. Regier and Hiebert [5] transferred individual fruits onto a cup conveyor belt, with each fruit landing on a cup, thus achieving fruit separation. However, multiple fruits may be placed on a single cup, affecting the acquisition of fruit surface information. Throop et al. [6] transferred apples using a ring wheel block with wheels at the bottom. The rings on the block hold the apples in place, and when the wheels contact the surface of the apples, friction drives the apples to rotate and thus collect full surface information. To improve efficiency, Throop added two small rotating axes on one side of the ring block that work in conjunction with the bottom wheel [7], and experimental results show that this new structure can rotate an apple within 6 s. The new structure is designed to be used in a variety of applications. In addition, this structure improves the stability of the apple by additional support from the two axes during transportation. Blasco et al. [8] developed a vacuum tube based apple rotation device. The device uses vacuum to adsorb the apples, and the rotation of the vacuum tube exposes the different surfaces of the apples to the camera field of view. Inner protruding roller is a commonly used ring-shaped apple transfer device invented by Hou et al. [9]. The roll located in the center of the ring is protruding and self-rotating, and the friction generated by its contact with the surface of the apples drives the apples to perform reverse rotation. The biconical roller conveyor is high at both ends and low in the center, with the apples located in the middle of the two rollers. As the individual rollers rotate simultaneously as they advance horizontally, the friction on the contact surfaces of the rollers and the apples drives the apples to rotate in the direction opposite to the direction of rotation of the rollers [10]. The earliest research on double cone rollers can be traced back to a patented invention in the 1990s [11], and with subsequent optimizations and improvements [12–14], double cone rollers have been the most widely used device for rotating apples in apple indoor grading systems. A roller conveyor consists of multiple rollers arranged perpendicular to the direction of apple conveyance, and the rotation of apples is achieved by friction between the rollers and the apples [15]. The difference between biconical rollers and roller conveyors is that there is a recessed area between the two biconical rollers to restrict the movement of the apples, whereas the apples are free to move between the two rollers of the roller conveyor. Cheng et al. [16] and Sadegaonkar and Wagh [17] designed a multi-channel roller conveyor to separate the apples, whereas Sofu et al. [18] added a roller brush to make the separation of the apples more effective.

The double cone roll is designed for indoor grading, but when the application scenario changes to the field, factors such as mechanical vibration, uneven ground and dust in the garden can seriously affect its performance [19]. For example, when the unit is tilted, gravity tends to dislodge the apples from the biconical roll. Since the vacuum tube needs to adsorb and rotate the apples to complete the whole surface information collection, it takes several seconds or even more than ten seconds to process a single apple, so it cannot meet the requirements of high throughput in the field. Fruit cup type transmission method requires a larger space, cannot be used in the field of smaller space. And the fruit cup needs more complex mechanical structure to work, in the field use situation, easy to fail; internal protruding roller device if the roller is too small, it will reduce the contact area with the apple, resulting in friction is not enough to drive the apple rotation. Too large a roller size, on the other hand, may result in the apples being positioned too high, reducing the contact area between the apples and the device and making it easier for the apples to fall out of the device. This makes it difficult to adapt the internal protruding roll design to the generalization of all practical production requirements due to individual differences in apple shape and the diversity of morphology between varieties.

Apple field grading technology requires real-time grading of post-harvest apples under field operating conditions such as using small space, uneven field ground, mechanical vibration, etc. [20]. Therefore, apple field grading equipment is small in size and high in applicability, which can significantly reduce post-production losses and lower apple production costs. First, during the field picking process, apples of different qualities are mixed together and stored in the cold, and the bad fruits are

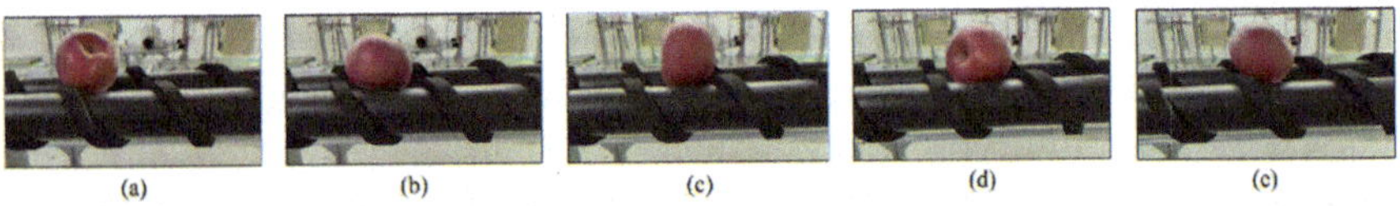

Fig. 1 Periodic horizontal acceleration and deceleration of an apple on a conveyor rod: **a** the thread touches the apple and pushes the apple to accelerate, **b** the apple leaves the thread, **c** the apple is farthest away from the thread, **d** the distance between the apple and the thread decreases gradually, and **e** the apple touches the thread again and accelerates

easy to mold and rot in the storage process, thus infecting the other fruits, resulting in the whole batch of apples not being able to enter the market circulation, which leads to a large economic loss. Secondly, the field picked apples of different grades, due to post-production treatment grading production line cost is higher, if most of the fruit is sold at a lower price of the second fruit, will lead to the cost of post-production grading of the farmers higher than the sale of earnings. Therefore, China's apple industry urgently needs to develop apple field grading equipment to improve the market competitiveness of China's apples [21].

A screw conveyor is a type of conveyor that transports and rotates apples, using threads on rods to drive the apples forward and rotate them. Zhang et al. [22] designed an apple field grading rig that uses three pairs of variable-pitch screw conveyor rods (the pitch increases with the direction in which the apples are moving) to separate and rotate the apples. The threads were encased in a sponge to prevent bruising of the apples during transportation. The threads push the apples to create acceleration, causing the apples to accelerate (Fig. 1a). Once the apples move horizontally at a speed greater than the horizontal speed of the threads (Fig. 1b), the apples are no longer pushed by the threads (Fig. 1c), and the apples then decelerate (Fig. 1d), then the apples come into contact with the threads (Fig. 1e), and so on in a cyclical acceleration and deceleration of the apples.

Apple size affects the trajectory of apple, the larger the apple the greater its mass, the greater the inertia, and the same collision force produces different acceleration; the speed of the conveyor rod determines the frequency of collision between the thread and the apple, which affects the speed law of the apple's horizontal motion; the size of the pitch determines the magnitude and direction of the collision force between the thread and the apple and the apple's horizontal motion distance. Therefore, this study aims to evaluate the performance of the separating rotating mechanism in the apple field grading equipment, to investigate the relationship between apple size, conveyor rod rotational speed, pitch and apple rotation and horizontal motion, and to study the rotational motion law and horizontal motion distance of apples on the basis of which the theoretical basis is provided for obtaining the full-surface information of apples, and realizing the high-efficiency grading of apples in the field. The specific objectives of the study are:

- Development of a color-based surface segmentation algorithm for motion apples to determine the surface area of image samples;

- To study the relationship between apple size, conveyor rod rotation speed and apple movement, and to verify the versatility of the separation rotation mechanism;
- To study the relationship between pitch size and horizontal movement of apples, to determine the horizontal movement distance of apples, and to realize accurate grading;
- Evaluate the rotational movement pattern of the apple and determine the number of apple rotations to ensure that full-surface apple information acquisition is achieved.

2 Materials and Methods

2.1 Apple In-field Sorting Equipment

The apple in-field sorting equipment based on machine vision consists of three main parts: The singulating and rotating mechanism (SRM), the computerized imaging system and the sorting actuator. Figure 2 shows the schematic diagram of the apple in-field sorting equipment.

The separation rotation mechanism (Fig. 3) consists of three single rows of channels, each with a pair of screw drives, which rotate, separate, and transfer the picked apples to the imaging chamber. The screw drivers consisted of long cylindrical aluminum tubes filled with 10 mm foam rubber tubing insulation. A self-adhesive sponge tape (15 mm high, 15 mm wide) was used to wrap the center tube to form the spiral shape with a thread pitch of 80–120 mm. the variable pitch design allows for efficient separation of apples moving on the conveyor. As the apples move between small to large thread pitches, the apples are separated by increasing their linear speed.

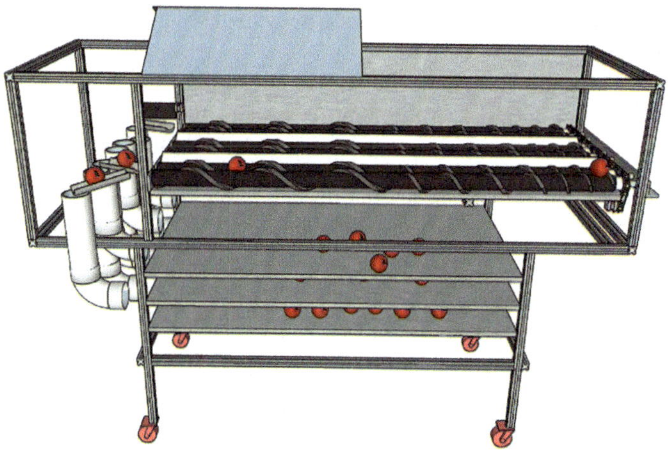

Fig. 2 Schematic diagram of apple in-field sorting equipment

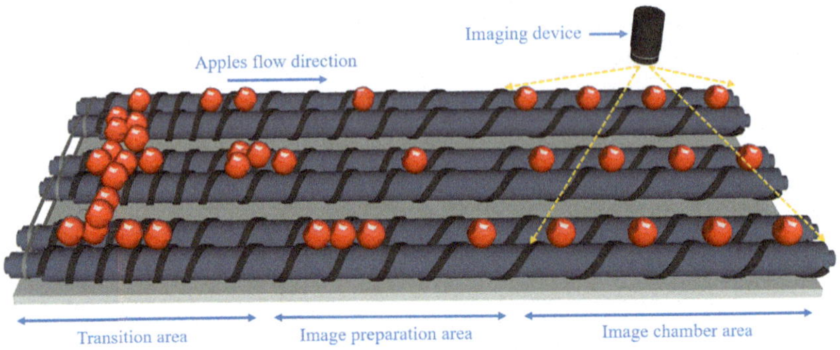

Fig. 3 Schematic diagram of separation rotating mechanism (SRM)

The rotation of the apples is achieved by a screw drive and a foam-filled cylindrical shaft. The separation rotary mechanism tested in this study operated at a motor speed of 2–3 rpm and a sorting actuator was built to orient the graded apples to the target bins.

A CCD color camera (Hikvision MV-CB013-A0UM-C) was mounted 530 mm above the imaging area for image acquisition, covering three channels of the SRM. It has a field of view (FOV) area of approximately 612 × 525 mm, and the color images were acquired at 12 fps with a resolution of 1440 × 1080 pixels. 2 tubular LEDs, 12 W each, were used to provide uniform illumination. Fruit grading was performed by an automatic adjustable segmentation algorithm for color images based on linear support vector machine (SVM) and Otsu thresholding method [23], which avoids grading errors due to variations in lighting conditions and fruit colors. Yu et al. [24] designed a maximum cross-section diameter detection method based on the orientation of apples, which takes the maximum diameter perpendicular to the orientation of apples its as the maximum cross-section diameter, and accurately detects the apple fruit diameter with an accuracy of 100%, which provides technical support for the size grading of apples.

After the computer system processes and analyzes the obtained images, the computer will send a signal to the control system, the sorting actuator will sort the apples according to the grading results, and the paddle sorting actuator will guide the apples to different grade bins.

2.2 Sample Preparation

The separation rotation mechanism should be tested and evaluated in both laboratory and orchard environments to ensure that the apple field grading equipment is capable of meeting field grading requirements. Specifically, the conveyor bar should complete a stable and consistent rotation of the apples at the maximum allowable speed for the three lanes, separating all apples one by one and obtaining full-surface information

Fig. 4 Six colored apples (2 each of large, medium, and small apples)

about the apples. This study focused on testing and evaluating the separation rotation mechanism in the laboratory only.

The most common apple on the market is the "Red Fuji" apple. The size of the apples and the conveying speed as well as the pitch length of the conveyor rod are the main factors affecting the rotation of the apples on the conveyor rod. Therefore, three different sizes of apples (large, medium, and small) were tested at two conveyor speeds (2 rpm and 3 rpm motor) with average fruit diameters of 96.9 cm (large), 86.4 cm (medium), and 74.6 cm (small), respectively. There were 2 apples of each size for a total of 6 apples. Each apple was divided into six equal parts and six colors (red, white, yellow, blue, green, and purple) of watercolor paint were selected and evenly applied to the apples so that the paint completely covered the original skin color of the apples, which was used to better observe the surface area of the apples under the imaging chamber (Fig. 4).

2.3 *Image Acquisition and Processing*

Figure 5 shows a flowchart of the acquisition and processing of the areas of the six color regions in the acquired apple motion image as a means of evaluating the performance of the separation rotation mechanism in apple delivery.

2.3.1 Image Acquisition

The conveyor rod motors were operated at 2 r/s and 3 r/s, respectively, and the six apples were conveyed five times sequentially in the separation, preparation, and imaging zones, respectively, and a camera was set up sequentially above the three zones to acquire the color image samples at a speed of 12 frames per second. A CCD

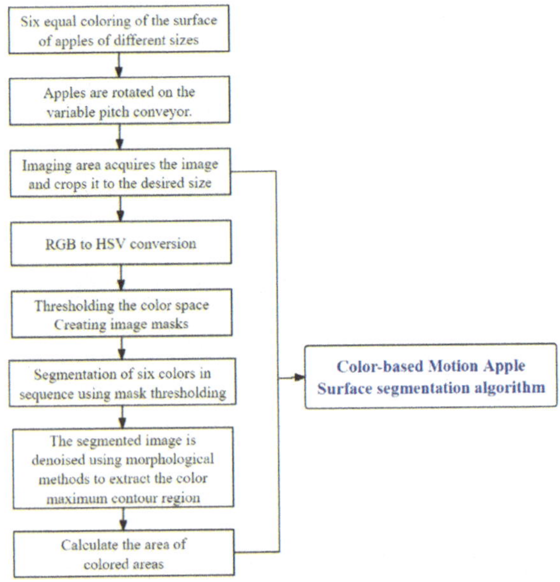

Fig. 5 Flow chart of color region acquisition and processing of apple motion image

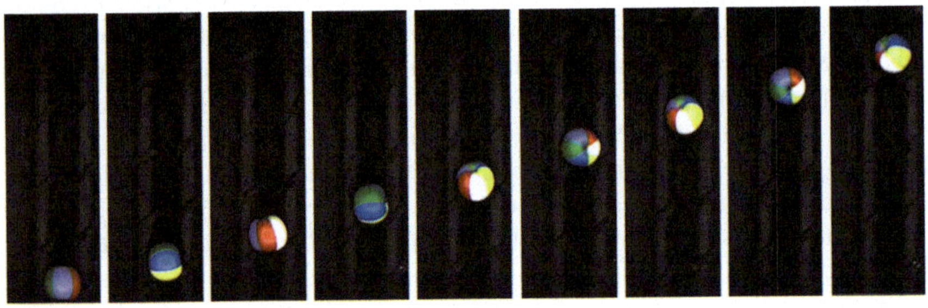

Fig. 6 Sample image obtained after cropping

color camera (Hikvision MV-CB013-A0UM-C) was used to acquire the raw image samples with a camera field of view of approximately 1440 × 1080 mm, covering the three channels of the separation rotation mechanism. The region of interest covering only the middle channel was obtained by cropping the original image to 360 × 1080 pixel size (Fig. 6). The number of images obtained for each apple varies with the transmission speed since the conveyor operates at two different speeds.

2.3.2 Image Segmentation

The HSV color space is close to people's perceptual experience of color, and can very intuitively express the hue, vividness, and lightness and darkness of colors (Fig. 7). Compared with the RGB color space, the HSV model has better color evaluation performance, so all cropped images were converted from the RGB model to the HSV model. Figure 8 shows the color histogram of six colors extracted from an apple sample. It can be seen that the hue value (Hue), saturation (Saturation) and brightness (Value) of the six colors differ significantly. The six color spaces are thresholded, an image mask is created, and the six colors are segmented sequentially using mask thresholding. After the initial segmentation of the image will produce some additional small areas of color points or regions, using the morphological processing method of small area region removal, the image can be denoised, and finally clearly segmented out of the six colors to form the desired color region (Fig. 8).

The color-based motion apple surface segmentation algorithm is written by MATLAB software to process and analyze the apple motion images by automatically detecting the coloring regions in the acquired images.

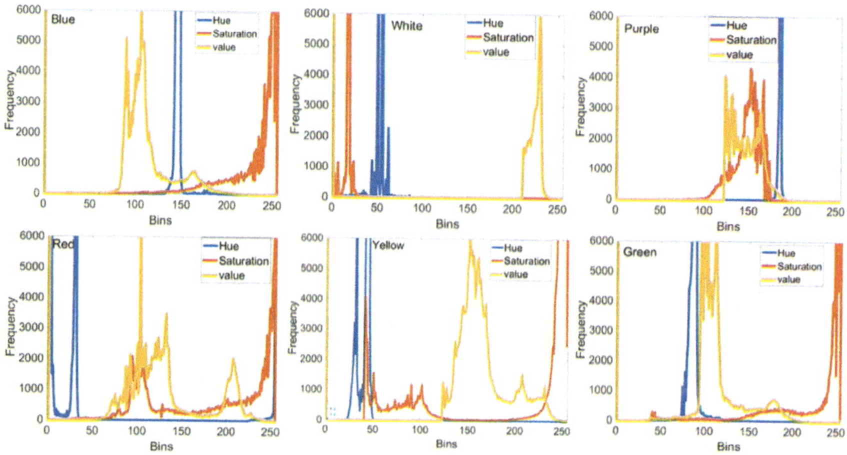

Fig. 7 Histogram of six colors

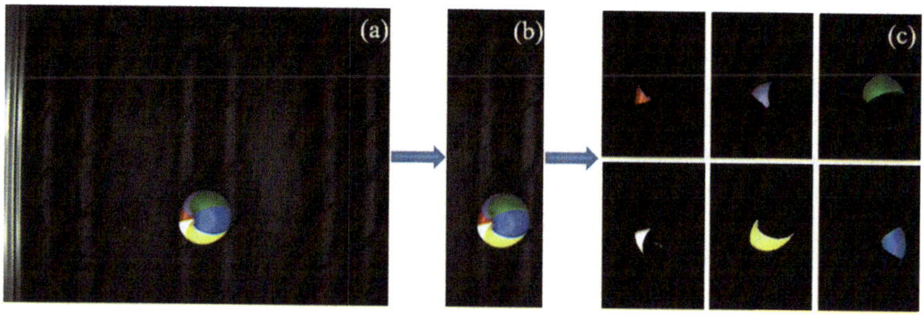

Fig. 8 **a** Original image, **b** cropped image and **c** segmented image

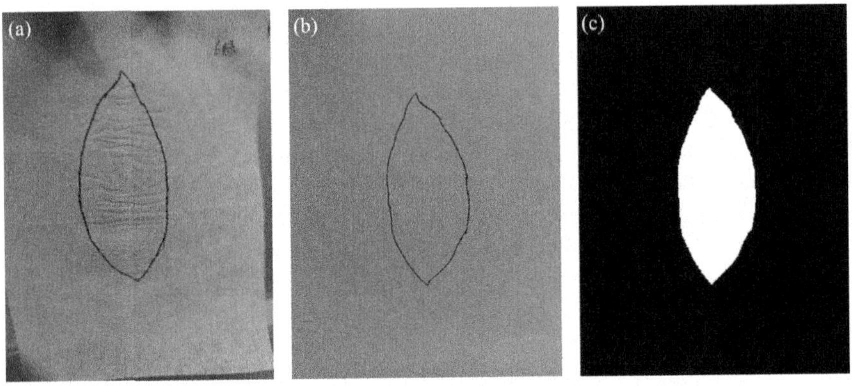

Fig. 9 The process of extracting the hand-drawn outline lines of the apple: **a** Capture image, **b** trace image and **c** segment image

2.3.3 Measurement of Actual Coloring Area of Apples

Since the entire surface of the apple is curved and the spatial resolution of the image varies with the position of the apple relative to the camera, the actual area of coloring on the apple needs to be measured in pixels. For this purpose, the actual area of all six colors of all six apples was measured using A4 paper.

First, cover each color area of the apple with a soft tissue paper, and then trace the boundaries of each color area on the tissue paper with a black marker. Since the tissue paper was soft and not easy to store, the outlines on the tissue paper were transferred to white paper by tracing, at which point a total of 36 sheets of white paper with different hand-drawn outlines were obtained. All the outlines were placed in the center of the transfer bar, and the camera was moved and adjusted so that the camera was positioned directly above the white paper and captured the image, ensuring that the height of the camera was the same as the height of the original apple movement to obtain the image in each area (Fig. 9).

Figure 10 shows the image processing process for extracting the hand-drawn lines of the apple contour. The traced image is transferred to a conveyor for capturing, followed by preprocessing of the image with graying and cropping to extract the hand-drawn lines and morphological closure operations, and finally the connected branch algorithm is applied to extract the central white block by traversing each pixel point of the image to divide the pixels in the image into a number of connected regions. During the implementation of the algorithm, either depth-first search (DFS) or breadth-first search (BFS) can be used for the traversal of pixel points and the construction of connected regions. The contour area measured in pixels in Fig. 10f represents the actual coloring area of the apple.

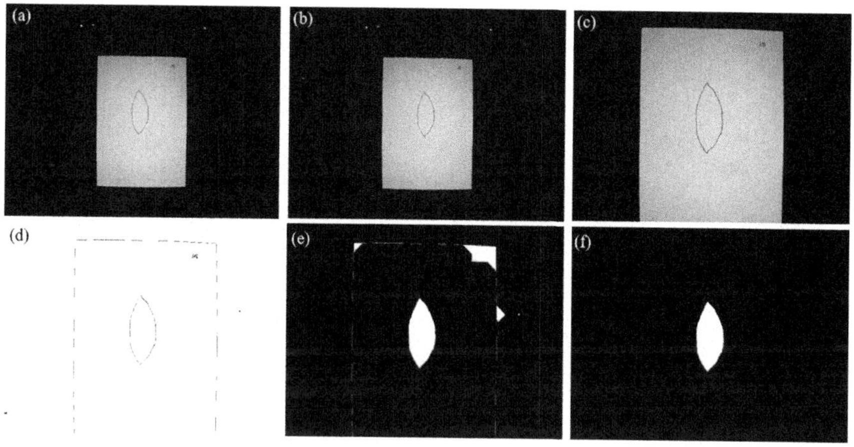

Fig. 10 Image processing for extracting hand-drawn contour lines of apple: **a** original image, **b** gray processing, **c** crop image, **d** extract hand-drawn line, **e** morphological closed operation and **f** the connected branch algorithm extracts the central white block

3 Results and Discussions

3.1 Actual Coloring Area of Apples

Due to the different heights of the cameras in the three areas during data acquisition, the pixel values obtained for the same image in different areas are different, which makes the amount of data cumbersome and redundant and is not conducive to the subsequent comparison of results. The pixel values obtained in the three areas are linearized with the camera height values to facilitate subsequent data processing.

The actual area of each color obtained from the measurements is shown in Fig. 11. Each of the horizontal axes of the bars represents the average pixel value of each particular color and the total value of the six colors superimposed on each other for the three sizes of apples. For different sizes of apples, both the area of each color and the total area increase linearly with increasing apple diameter; the discounting in Fig. 12 shows a comparison of the areas of the six colors for the same size of apple. For apples of the same size, their six color areas are linearly distributed and are essentially the same. The results of this data enhance the reliability of the apple rotation pattern analysis in this paper.

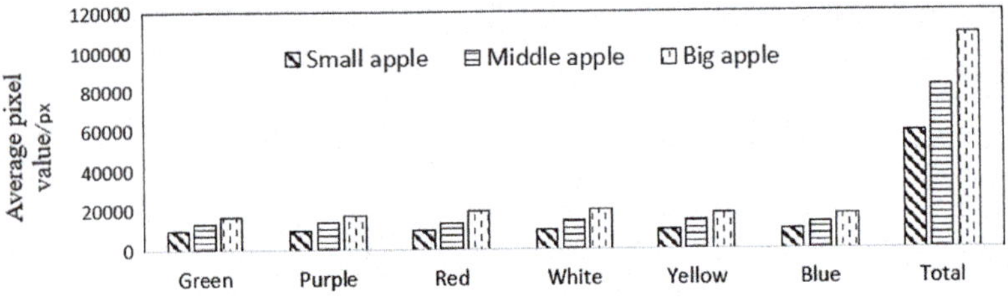

Fig. 11 Actual coloring area statistics of apples

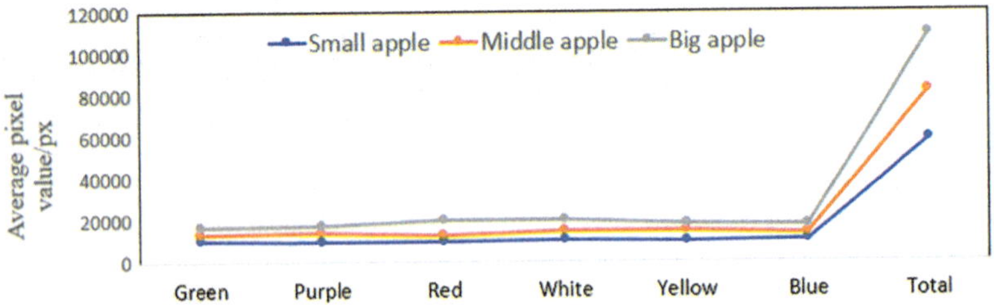

Fig. 12 Line graph of actual coloring area of apples

3.2 Relationship Between the Image Obtained in the Imaging Area and the Transmission Speed

Throughout the process of the apple on the conveyor, the acquisition of apple images starts only when the apple is intact and within the visual range of the camera. At the same transfer speed, the number of images acquired varies for apples of different fruit sizes, and the relationship between the acquired images and the transfer speed is characterized by calculating the mean and standard deviation of the images acquired for each apple.

At motor gear speeds of 2 rpm and 3 rpm, the camera acquired images at 12 frames/sec and the average number of apple motion images obtained with standard deviation is shown in Table 1. As the motor gear speed increases, the number of images acquired per apple decreases from 13 to 9. The faster the apples are transported, the fewer images are captured by the camera.

Table 1 Mean and standard deviation of images collected from apples at different speeds

Speed (RPS)	Images collected for each apple (mean/standard deviation)
2	13.28/0.35
3	9.69/0.29

3.3 *Average Cumulative Percentage of Apples*

A good apple rotation conveyor system should be able to rotate each apple in an appropriate manner to ensure that the imaging device is able to obtain full surface information for each apple, i.e., at least 100% of the cumulative area, while maintaining the inspection efficiency of the system. In addition, because each apple is rotated multiple times during imaging, it is important to ensure consistent imaging of each surface area to provide a more accurate and consistent assessment of fruit quality grades.

Cumulative percentage is a commonly used statistical indicator in statistical analysis to describe the proportion of the cumulative value of a variable over a period of time or at a point in time. The formula for calculating the cumulative percentage is:

$$\text{Cumulative percentage} = (\text{cumulative sum/total}) * 100\%$$

Cumulative Percentage is an intuitive and useful tool to help us better interpret and understand data, and quickly identify trends and patterns in a data set. For example, in a control chart, the cumulative percentage can help identify anomalies in the production process.

In the rotational pattern analysis of the apple, the cumulative percentage represents the percentage ratio of the sum of the pixel values of each color in all frames acquired by the apple to the total area of the apple actually colored. Its calculation formula is:

$$\begin{aligned}\text{Cumulative percentage} = {} & (\text{sum of individual color pixel} \\ & \text{values in all frames of the apple} \\ & /\text{total area of the apple actually tinted}) * 100\%\end{aligned}$$

At the time of acquiring the images, each apple was repeated five times on the conveyor, so an average cumulative percentage (Average cumulative percentage, ACP) was defined to average the cumulative percentages of apples of the same size and present the data in a histogram. The ACP shows the cumulative effect over time, visualizes the number of turns the apple made on the conveyor, and indicates whether the entire apple was rotated in a consistent pattern and indicates whether the entire apple is rotated consistently in the desired pattern. The calculated average cumulative percentage of an apple is shown in Table 2.

Table 2 Average cumulative percentage statistics for apples

Blue	Green	Purple	Red	White	Yellow	All
147,006	165,983	166,268	159,767	153,376	177,643	970,043
293,940	293,940	293,940	293,940	293,940	293,940	293,940
50%	56%	57%	54%	52%	60%	330%

3.4 Horizontal Movement Patterns of Apples

The relationship between the rotational motion of the apples and the pitch was verified in terms of three zones. The variable pitch spiral conveyor of the conveyor had a pitch that increased gradually with the movement of the apples, where the pitch of the two neighboring spiral axes in the separation, preparation, and imaging zones were 8 cm, 10 cm, and 12 cm, respectively. Figure 13 shows the ACPs of the apples in the three zones reflecting the number of revolutions of the apples at different rotational speeds within the separation, preparation, and imaging zones. Since the three zones are not divided by the same length, with the separation zone being 50.0 cm long, the preparation zone being 40.0 cm long, and the imaging zone being 60.0 cm long, the number of rotational revolutions combined with the length of the zones can be used to calculate the distance traveled by the apple in one rotation. By combining the time taken by the apple to travel through the entire zone, the time taken by the apple to rotate one revolution and the distance traveled by the apple per second can be calculated, as shown in Tables 3 and 4.

The results show that the time taken to rotate an apple one turn remains essentially the same at 0.8 s for different pitch conveyors and does not vary with the speed of

Fig. 13 Average cumulative percentage statistics for apples in three regions

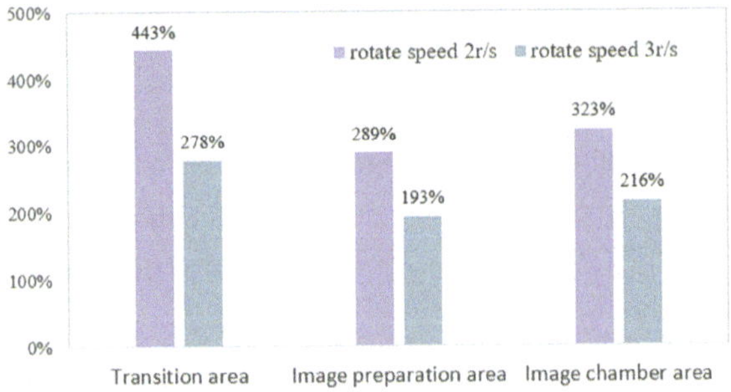

Table 3 Horizontal movement of apples at 2 r/s conveyor rod speed

Rotate speed: 2 r/s			
	Transition area	Preparation area	Chamber area
Length (cm)	50.0	40.0	60.0
Number of turns	4.4	2.9	3.2
Time for one revolution (s)	0.7	0.8	0.8
Time taken for apples to pass through the region (s)	3.2	2.4	2.6
Distance travelled by an apple per second (cm)	15.6	16.7	23.1
Distance travelled by the apple in one revolution (cm)	11.4	13.7	18.6

Table 4 Horizontal movement of apples at 3 r/s conveyor rod speed

Rotate speed: 3 r/s			
	Transition area	Preparation area	Chamber area
Length (cm)	50.0	40.0	60.0
Number of turns	2.8	1.9	2.2
Time for one revolution (s)	0.8	0.8	0.8
Time taken for apples to pass through the region (s)	2.2	1.6	1.8
Distance travelled by an apple per second (cm)	22.7	25.0	33.3
Distance travelled by the apple in one revolution (cm)	17.9	21.1	27.3

the conveyor bar. A single rotation time of less than 1.0 s verifies the timeliness and stability of the conveyor. In the imaging area, the camera can cover four thread pitches in three conveyor lanes, i.e., a single photo of the imaging area can show the rotating images of 12 apples at the same time, so the grader can be efficient enough to process the surface information of 12 apples at the same time and grade them in less than 1.0 s.

Tables 3 and 4 visualizes the distance per second traveled by the apple as it passes sequentially through the regions of different pitches at 2 and 3 r/s, as well as the distance traveled in one rotation. It can be seen that as the pitch increases, the distance traveled by the apple in one rotation increases. The change in this result is mainly due to the fact that when the pitch of the conveyor belt is small, the apple is pushed forward by the previous thread in the thread gap, and during the acceleration process, it hits the latter thread and backs up for a certain distance, which results in a shorter distance traveled by the apple. During the backward process, part of the surface image of the apple is repeatedly captured, so at a small pitch, the distance traveled by the apple per second is shorter, and the distance used to rotate a circle is also shorter. As the pitch increases, within the imaging area, the apple is not hit by the latter thread during forward acceleration, and the apple undergoes a more consistent and smooth rotational motion with a large pitch, and its surface image information is captured more efficiently. Therefore, while ensuring that the whole surface information of the apple can be obtained in the imaging area, the larger pitch makes the horizontal motion of the apple more consistent and smooth after separation, and it is more efficient and reliable in acquiring the image information, which provides a good condition for the next step of the computer system to classify the apple's size, color, and the degree of defects, etc. through the image.

3.5 Apple's Rotary Motion Pattern

Understanding the rotational motion patterns of the apples is important for the imaging system to acquire full surface information of the apples to ensure that the imaging system has captured the complete surface information of the apples at least once so that it can be effectively graded. Using samples of apple rotational images acquired under the imaging zone at two speeds, the rotational motion patterns of the apples on the conveyor bar are discussed by calculating the ACP of each apple. The pixel values of six colors for 13 sample images of an apple at a rotational speed of 2 r/s were recorded as in Fig. 14, and the pixel values of six colors for 9 sample images of an apple at a rotational speed of 3 r/s were recorded in Fig. 15. Each colored line in both figures follows a wave-like pattern, and each colored wave line has three peaks and two peaks at rotational speeds of 2 r/s and 3 r/s, respectively, which indicates that the apple on the conveyor rod undergoes a continuous and consistent rotation driven by the threads, and the number of apple rotations is not less than one revolution in each case.

Figure 16 counts the average cumulative percentage of three sizes of apples (large, medium, and small) at two speeds (2, 3 r/s). When the conveyor bar was operated at 2 r/s, the average cumulative percentages of large, medium, and small apples were 317%, 323%, and 330%, respectively, and the entire surface of each apple was imaged more than three times (300% or more); when the conveyor bar was operated at 3 r/s, the average cumulative percentages of large, medium, and small apples were 208%, 216%, and 219%, respectively, and the entire surface of each apple was imaged more than two times (200% or more). Entire surface was imaged more than 2 times (200% or more). It can be seen that the number of apple rotations decreases as the rotational speed increases, but it ensures that the imaging system captures the complete surface information of the apple. The ACP distribution of each of the six colors at both

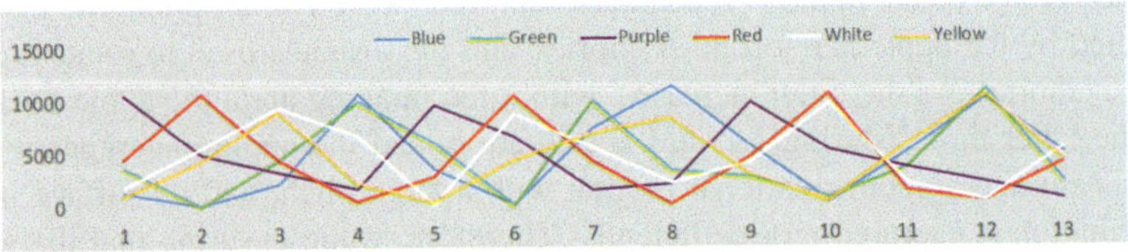

Fig. 14 Line graph of pixel values for each color in 13 sample images at 2 r/s

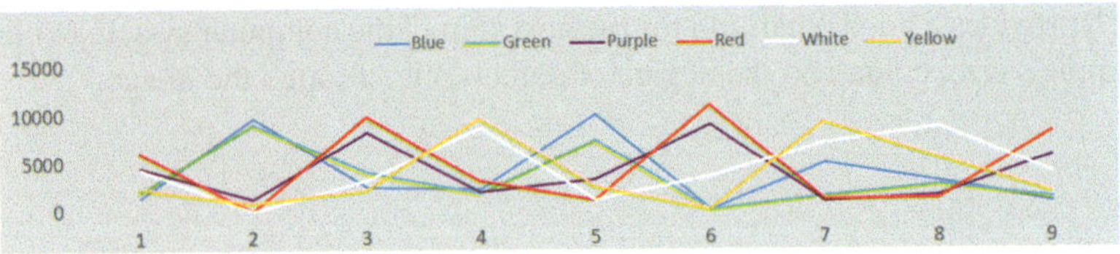

Fig. 15 Line graph of pixel values for each color in 9 sample maps at sorting speed 3 r/s

speeds was also more even, between 50–60% and 30–40%, indicating that the image samples of the six colors appeared to be more uniform, and that the apples could be rotated consistently and stably, providing the imaging system with effective apple surface information. By comparing the number of revolutions of large, medium and small apples at the same speed, it was found that the number of revolutions of apples was approximately the same for different sizes of apples, and that the apple size did not have a significant effect on the change of ACP, which indicated that the foam covering the helix shaft effectively rotated the apples with little slippage.

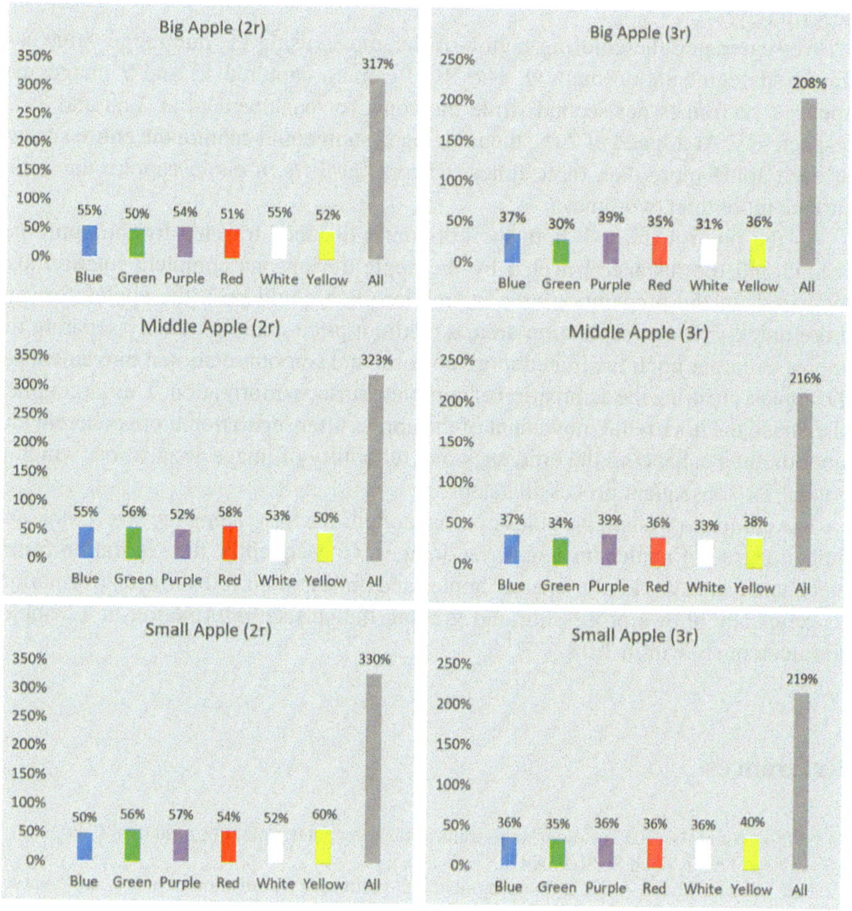

Fig. 16 ACP statistics for three sizes of apples (large, medium and small) at two speeds (2, 3 r/s)

4 Conclusions

In this study, we utilized the separation rotary mechanism in laboratory-developed apple in-field sorting equipment as a focal point. We developed a color-based surface segmentation algorithm for moving apples and investigated their rotational and horizontal motion behaviors on a variable-pitch conveyor rod at two conveyor speeds. Additionally, we examined the imaging capabilities and clarified the relationship between apple size, conveyor rod rotation speed, and pitch size concerning the movement of apples. The performance of the variable-pitch conveyor device was evaluated accordingly.

We segmented the coloring region of the apples using the hue range value as a threshold segmentation method. The RGB camera captured 13 and 9 images per apple at 12 frames per second while the conveyor bar operated at 2 r/s and 3 r/s, respectively. At a speed of 2 r/s, the imaging system could capture the entire surface of each apple more than three times, whereas at 3 r/s, it could capture the entire surface more than two times.

As the pitch increased, both the horizontal distance traveled by the apple per second and the distance traveled by the apple during one complete rotation also increased. In the preparation area, a smaller pitch could push the apples forward more quickly; in the separation area, a medium pitch could gradually separate the apples. A larger pitch facilitated more uniform and smooth rotational movements of the apples, ensuring the acquisition of complete surface information. This larger pitch also made the horizontal movement of the apples after separation more coherent and smooth, thus enhancing the efficiency and reliability of image acquisition, which is crucial for subsequent processing steps.

The computer system provides optimal conditions for grading the size, color, and defect degree of apples through these images. Consequently, the separation rotary mechanism can efficiently separate apples and present their full surface information. Its efficiency allows processing and grading the surface information of 12 apples simultaneously within 1.0 s.

References

1. Wu D, Sun DW (2013) Colour measurements by computer vision for food quality control—a review. Trends Food Sci Technol 29(1):5–20
2. Lu Y, Lu R, Zhang Z (2022) Development and preliminary evaluation of a new apple harvest assist and in-field sorting machine. Appl Eng Agric 38(1):23–35
3. Pothula AK, Zhang Z, Lu R (2023) Evaluation of a new apple in-field sorting system for fruit singulation, rotation and imaging. Comput Electron Agric 208:107789
4. Suzuki M, Maeda H (1989) Method and apparatus for inspecting form, size and surface condition of conveyed articles by reflecting images of four United States Patent 4825068
5. Regier RD, Hiebert JF (1993) U.S. Patent No. 5244100. Washington, DC: U.S. Patent Office. Patent and Trademark Office
6. Throop J, Aneshansley D, Anger W, Peterson D (2003) Conveyor design for apple orientation. Paper presented at the 2003 ASAE annual meeting

7. Throop JA, Aneshansley DJ, Anger WC, Peterson DL (2005) Quality evaluation of apples based on surface defects: development of an automated inspection system. Postharv Biol Technol 36(3):281–290
8. Blasco J, Aleixos N, Moltó E (2003) Machine vision system for automatic quality grading of fruit. Biosyst Eng 85(4):415–423
9. Hou WJ (2006) Research on automatic apple grading method based on machine vision [Master's thesis]. Nanjing Forestry University, Nanjing
10. Throop JA, Aneshansley DJ, Anger WC, Peterson DL (2003) Conveyor design for apple orientation. ASABE Paper No. 036123, St. Joseph, MI: ASABE
11. Hiebert JF (1996) Method and apparatus for handling objects. U.S. Patent No. 5497887
12. Hiebert JF (1997) Method and apparatus for handling objects. U.S. Patent No. 5626236
13. Mizushima A, Lu R (2013) A low-cost color vision system for automatic estimation of apple fruit orientation and maximum equatorial diameter. Trans ASABE 56(3):813–827
14. Ying Y-B, Xu H-R, Ye Y-C (2002) Double cone type roller fruit conveying and turning device for grading machine. China Patent 02266030.5.2002
15. Fan S, Li J, Zhang Y, Tian X, Wang Q, He X, Huang W (2020) On line detection of defective apples using computer vision system combined with deep learning methods. J Food Eng 286:110102
16. Cheng X, Tao Y, Chen Y, Luo Y (2003) Nir/MIR dual-sensor machine vision system for online apple stem-end/calyx recognition. Trans ASAE 46(2):551
17. Sadegaonkar V, Wagh K (2015) Automatic sorting using computer vision & image processing for improving apple quality. Int J Innov Res Dev 4(1):543–546
18. Sofu M, Er O, Kayacan M, Cetişli B (2016) Design of an automatic apple sorting system using machine vision. Comput Electron Agric 127:395–405
19. Li J, Zhang D, Liu B (2009) Design of conveying and turning mechanism of apple grading machine. J Agric Mach 40(5):58–161, 157
20. Pothula AK, Zhang Z, Lu R (2018) Design features and bruise evaluation of an apple harvest and in-field presorting machine. Trans ASABE 61(3):1135–1144
21. Zhang Z, Igathinathane C, Li J, Cen H, Lu Y, Flores P (2020) Technology progress in mechanical harvest of fresh market apples. Comput Electron Agric 175:105606
22. Zhang Z, Lu Y, Lu R (2021) Development and evaluation of an apple infield grading and sorting system. Postharv Biol Technol 180:111588
23. Mizushima A, Lu R (2013) An image segmentation method for apple sorting and grading using support vector machine and Otsu's method. Comput Electron Agric 94:29–37
24. Yu J, Zhou L, Zhao Z et al (2024) Maximum cross-section diameter detection method based on apple orientation. J China Agric Univ 29(05):125–134

Research on the Transferability of SSC Detection Models Between Different Instruments for Fresh Apricots

Runrun Wang, Shujuan Zhang, and Zhao Zhang

Abstract The determination of soluble solid content serves as a crucial parameter for assessing the internal quality of fresh apricots, thereby enhancing the precision of online detection and ensuring its significance in improving the accuracy of measuring soluble solid content in fresh apricots. To investigate spectral standardization methods for different NIR hyperspectral spectrometers, minimize spectral variations, and enable model sharing, we focused on soluble solids as the detection index and examined the spectra of fresh apricot samples collected by two NIR hyperspectral spectrometers. We employed direct standardization, slope/intercept correction, and single linear regression standardization algorithms to conduct our research on spectral standardization. The continuous projection algorithm is employed to enhance the model's transmission capability by selecting the characteristic wavelength variable. The model integrates the single linear regression normalization algorithm with the preferred wavelength variable and can effectively be applied to spectral data collected using different instruments. Consequently, there is a significant improvement in the prediction set R_p^2 from 0.1541 to 0.9405 and a remarkable decrease in RMSEP from 1.6161 to 0.3409, which facilitates model sharing across diverse instrument platforms.

Keywords Fresh apricot · Soluble solid content · Model transfer · Standardized algorithm

R. Wang · S. Zhang (✉)
College of Agricultural Engineering, Shanxi Agricultural University, Shanxi 030801, China

Z. Zhang (✉)
College of Information and Electrical Engineering, China Agricultural University, Beijing 100083, China
e-mail: zhaozhangcau@cau.edu.cn

Y. Liu et al. (eds.), *New Technologies Applied in Apple Production*, Smart Agriculture 10, https://doi.org/10.1007/978-981-97-7778-5_7

1 Introduction

The fresh apricot, a member of the rose family [14], is a plant with a rich history spanning over 4000 years. It blooms in spring and bears fruit in summer; both its flowers and fruits have captivated writers throughout time with their exquisite beauty. The flesh of fresh apricots is characterized by its succulence, sweetness, and occasional slightly sour, while also boasting numerous health benefits for human consumption. On one hand, fresh apricots exhibit a laxative and intestinal stimulatory effect, thereby facilitating the normal functioning of the digestive system and maintaining intestinal health [18]. On the other hand, specific constituents found in fresh apricots may provide protection against cardiovascular disease. The soluble solid content is a crucial indicator for assessing the quality of apricots, including sugar, acid, vitamins, minerals, and other constituents. These components have the potential to influence the taste, flavor profile, and nutritional value of apricots. Therefore, the determination of soluble solids is crucial for assessing fruit maturity, facilitating graded sales, and conducting scientific research. Conventional techniques for detecting soluble solids in fruits often involve destructive methods such as pressing, the addition of chemical reagents, and other approaches. These methods not only consume time and labor but also result in sample wastage. The application of quantitative analysis of fruit spectra enables the non-destructive testing of internal qualities, such as sugar content and firmness [4, 10–12]. The technology exhibits the attributes of rapidity, precision, and non-invasiveness, thereby providing robust support for the advancement of the fruit industry.

Spectral technology, which enables the acquisition of spectral information from fruit samples, has gained significant traction in the quantitative and qualitative analysis of fruit quality, as well as in determining fruit varieties and tracing their origins. The principle of spectral technology is primarily based on the interaction between light and matter [22]. When light illuminates a substance, its constituent particles such as electrons and protons undergo transitions in their energy levels, resulting in the absorption, emission, or scattering of light. By analyzing variations in wavelength and intensity during these phenomena, valuable insights into the internal structure and properties of matter can be gleaned. Spectral technology offers the advantages of rapid and non-destructive analysis, making it a successful tool in assessing fruit quality. This innovative approach overcomes the limitations of traditional physical and chemical detection methods, such as sample destruction and labor-intensive procedures.

The near-infrared region of the electromagnetic spectrum offers significant advantages for detecting and analyzing fruit components. Yu et al. [27] utilized hyperspectral reflection imaging to detect the hardness and SSC of Korla Pear and developed a deep network comprising an automatic encoder and a fully connected neural network. By employing this method, the correlation coefficients between the hardness and SSC prediction set of Korla pear reached 0.890 and 0.921. Xu et al. [23] utilized hyperspectral and fluorescence spectroscopy techniques to accurately determine the soluble solid content of kiwifruit, and further optimized the model by integrating five pretreatment methods and two characteristic wavelength extraction approaches. As a result, the correlation coefficients for the prediction set were 0.8894 and 0.8717, respectively. Rahmia et al. [16] utilized visible near-infrared spectroscopy to investigate the variation in malic acid content with storage temperature. After establishing the model, the prediction models demonstrated correlation coefficients of 0.89, 0.87, and 0.92 for apple acid content at temperatures of 8 °C, 13 °C, and 29 °C respectively, showcasing the accurate capability of predicting malic acid content under different storage temperatures. To address the limitations of the spectral model in predicting new samples from different varieties or batches, Huang et al. [9] utilized a transfer learning approach using 9 batches of tea samples from 3 distinct varieties. This strategy significantly enhanced the robustness of the tea water spectral determination model by mitigating variations among different tea samples in their spectral profiles. These studies primarily focus on developing predictive models for one or more indicators using spectral devices. To the best of our knowledge, no investigations have been conducted on the transferability of detection models across different spectral instruments for determining fresh apricot SSC.

The universality, stability, and accuracy of the calibration model pose a significant challenge in current research [8, 20, 21]. The spectral response characteristics of the same sample may vary due to differences in instrumentation and measurement conditions, resulting in inconsistent detection outcomes across devices with varying resolutions, thus hindering the sharing of data resources. The direct correction, piecewise direct correction, and slope/bias algorithms are commonly used in model transmission; however, their effectiveness is relatively limited, which still imposes certain constraints on practical applications to some extent. To address the issue of non-shared calibration models across multiple channels due to spectral differences, Sun et al. [19] proposed a segmented cosine spectrum-based method for spectral standardization, effectively mitigating the impact of inter-channel spectral variations on PLS applicability. Therefore, investigating model transmission, enhancing and innovatively applying algorithms, and analyzing the interaction mechanism between light and biological tissues all contribute to establishing a more efficient and stable detection model for quality evaluation.

The aim of this study was to investigate the transferability of a fresh apricot SSC detection model between two different spectral instruments. The specific objectives were as follows: (1) Comparing the differences in spectral data obtained using two hyperspectral instruments; (2) Evaluating the effectiveness of three spectral standardization algorithms in correcting these differences; (3) Utilizing the SPA method to optimize wavelength variables for enhancing the model.

2 Materials and Methods

2.1 Sample Preparation and SSC Measurement

The "6–1" fresh apricot selected in the experiment is a new variety cultivated by the apricot base of the Fruit Research Institute at Shanxi Agricultural University, characterized by its firm texture and superior quality. The samples were collected prior to the fruit reaching full ripeness in July 2022, when the vibrant green color of the fresh apricot had significantly faded and the flesh had acquired a slightly softer texture. We carefully handpicked 295 ripe "6–1" apricots that exhibited a uniform shape, were devoid of any signs of pests or diseases, and demonstrated consistently high quality. During the selection process, a total of 295 fresh apricot fruits with a uniform shape, free from pests and diseases, and consistent quality were carefully chosen. Immediately after harvest, these fruits were transported to the laboratory where they underwent thorough cleaning before being stored in a controlled environment (15 °C) for 24 h. This step aimed to minimize any potential environmental influence on spectral acquisition and indicator measurement. After collecting spectral data, the apricots were systematically arranged and their Soluble Solid Content (SSC) was measured using a digital refractometer (SW-55S).

2.2 Spectral Data Acquisition

2.2.1 Near-Infrared Hyperspectral Imaging System

The HSI-1, a hyperspectral imager (900–1700 nm) developed by Beijing Zhuolli Hanguang Company, is designed for Gaia Sorter applications. The instrument primarily consists of a computer and an imaging spectrometer. The imaging spectrometer comprises the Image-N17E spectral camera, four 35W halogen lamps, a stepping-motor controlled mobile platform, and a camera obscura. All imaging systems are installed within the camera obscura. The distance between the lens and the sample is set to 220 mm before data collection. The exposure time is adjusted to 0.05 s while maintaining a platform running speed of 2.0 mm/s in order to enhance the stability of spectrometer data collection.

The HSI-2, a hyperspectral imager (900–1700 nm) developed by Headwall Photonics Company, is designed for Starter Kit applications. The system consists of an indoor mobile scanning platform, a miniature near-infrared hyperspectral imager (with a digital lens focal length of 25 mm and aperture of F/1.3), a PHILIPS 6995P light source model with a working color temperature of 3200 K and rated voltage/power of 230 V/1000W, and a resolution of 0.727 nm. Additionally, it includes a halogen light source, controller, and computer. The lens-to-sample distance is set at 220 mm, the platform velocity is 2.829 mm/s, and the exposure duration is 0.9 ms.

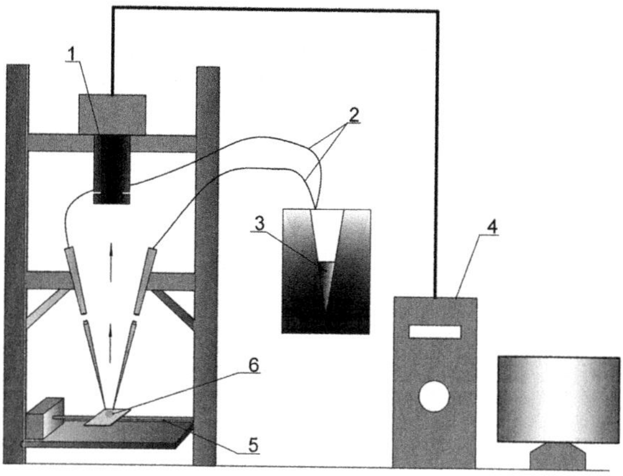

Fig. 1 Spectral imaging system 1—spectral camera, 2—optical fiber, 3—light source, 4—computer, 5—conveyer platform, 6—samples

2.2.2 Spectral Data Acquisition

The structure of the spectrum acquisition system is depicted in Fig. 1. The fresh apricot is placed on the mobile platform, and then the computer is used to activate the motor that propels the mobile platform. During the HSI process, the spectrometer camera scans the moving fresh apricot and captures its cartographic information. All collected data are subsequently recorded by the computer.

The hyperspectral image of the collected sample is subject to variations in light source intensity across different bands and dark current in the camera [3]. Therefore, it is essential to conduct black and white correction on the acquired hyperspectral image prior to data analysis and processing. The correction formula is presented in (1):

$$R = \frac{I - B}{W - B} \tag{1}$$

where R is the image data after black and white correction, I is the original image data, B is the blackboard image data, and W is the whiteboard image data.

2.3 *Data Analysis*

2.3.1 Model Transfer

The detection model based on the sample spectrum collected from one spectrometer is often ineffective for other types of spectral equipment [10–12, 20, 21]. The modeling of transmission has emerged as a significant obstacle in the practical implementation of spectroscopy, aiming to mitigate the impact caused by various instruments on spectral analysis. The spectral information of fresh apricot samples was

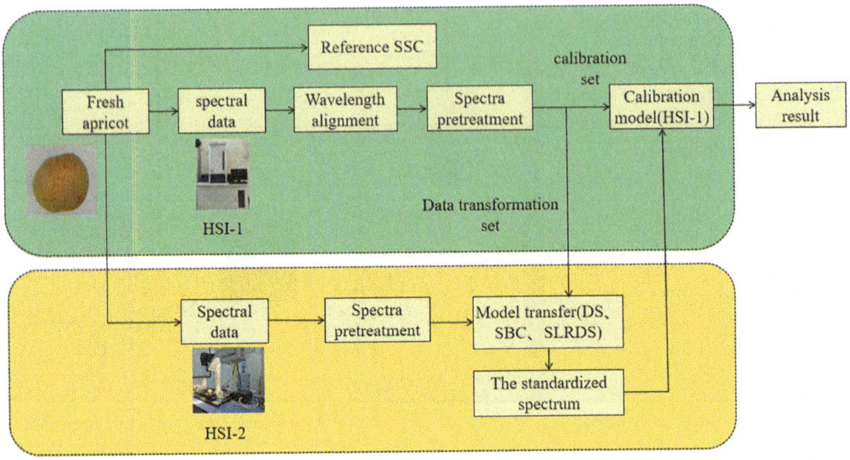

Fig. 2 The flow chart illustrates the process of transferring models between two spectrometers

collected using two hyperspectral instruments, and the disparities between the data acquired by different NIR hyperspectral instruments were analyzed. Subsequently, wavelength alignment was performed on the collected spectral data. The premise of spectral standardization and model transmission assumes that the spectra collected by different instruments should have an identical wavelength range and number, thus requiring the adjustment of instrument-collected spectral bands [2]. The spectral normalization algorithm was employed to establish a transfer matrix between the spectral data acquired by the two devices, and subsequently, the spectral data of HSI-2 were corrected. The Euclidean distance metric was utilized to assess both the degree of fitting for the spectral data and the effectiveness of standardization. The corrected HSI-2 spectral data was ultimately used in the HSI-1 calibration model to facilitate the sharing of the detection model between instruments. The process of model transfer is illustrated in Fig. 2.

2.3.2 Spectral Standardization Method

The purpose of model transfer is to facilitate the sharing of models and optimize resource allocation by reducing the need for redundant modeling efforts [5]. The compensation of wavelength distortion between different instruments is essential, as well as the correction of variations in reflected light received by each instrument caused by factors such as distance, light source, detector, and optical element. The offset between each slave and master instrument can be corrected by designing a standardized model for each individual instrument. The response measured at the exact wavenumber on the main instrument is found to be correlated with the response within a narrow range around each of the same wavenumbers measured from the instrument. The spectral response of the main device is linearly regressed with the values from the corresponding window on the device at each wave number, resulting in generating normalized coefficients for all wave numbers. The obtained spectrum can be normalized to the main instrument style by using a standardization coefficient whenever a new sample is measured, thus addressing the issue of limited generality in the spectral model.

The Direct Standardization (DS) algorithm [13] is based on the assumption that an instrument's influence can be characterized by quantifying the disparity between the host and slave spectra, and subsequently eliminating HSI influence from the instrument.

The technique of Slope/Bias Correction (SBC) is particularly employed in the context of data conversion or model transfer between instruments [6]. The primary objective is to enhance the consistency and reliability of the data by adjusting the forecast results obtained from the subordinate instrument in order to closely align them with those derived from the superior instrument. The fundamental principle of the SBC algorithm is to utilize the unitary linear relationship between the primary instrument and the secondary instrument for predicting results. The relationship typically manifests as a linear correlation, where the slope and intercept indicate the difference between the predicted outcomes of the two instruments [1]. The prediction from the slave instrument can be adjusted to align more closely with the prediction from the master instrument by calculating and incorporating the differences in slope and intercept.

The Single Linear Regression Normalization (SLRDS) algorithm considers the reflectance between each wavelength point as independent and applies a single linear regression to correct the spectrum. The SLRDS algorithm is implemented through the following steps.

The reflectance of the i sample at the j wavelength point is $S_m(i,j)$ and $S_t(i,j)$ respectively, which satisfy Eq. (2):

$$S_m(i,j) = b_0(j) + b(j)S_t(i,j) = \left[1\, S_t(i,j)\right] \cdot \binom{b_0(j)}{b(j)} \tag{2}$$

where $b_0(j)$, $b(j)$ are the regression coefficients of the spectral matrix of the master and slave instruments respectively The regression coefficient is calculated by the least squares of Eq. (3):

$$\binom{b_0(j)}{b(j)} = \left[1_{n\times 1}\, S_t(i,j)\right]^+ \cdot S_m(i,j) \tag{3}$$

After calculating the spectral matrix S_{unknow} of the unknown sample measured on the instrument through Eq. (4), the corrected spectral matrix of the slave machine is obtained:

$$S_{s,std} = \left[1_{n\times 1}\, S_{unknow}(i,j)\right] \cdot \binom{b_0(j)}{b(j)} \tag{4}$$

where S_{unknow} represents the spectral matrix of unknown samples measured on other instruments, and $S_{s,std}$ represents for the corrected slave spectral matrix. The corrected spectral matrix $S_{s,std}$ is brought into the model of the main instrument to obtain the predicted value after spectral standardization.

2.3.3 Wavelength Extraction

The wavelength should be filtered when establishing the NIR spectral model. The selection of wavelengths serves to streamline the model and, more importantly, facilitates the elimination of irrelevant or nonlinear variables [15]. The spectral wavelengths exhibit multiple correlations, specifically a linear correlation between wavelength variables, resulting in redundant information within the spectral data. This complexity complicates model calculations and diminishes prediction accuracy.

The Sequential Projection Algorithm (SPA) is a forward selection technique used for variable selection in multiple linear regression, principal component analysis, and other statistical modeling methods to address collinearity among the selected variables [24]. The main idea behind this algorithm is to identify the combination of variables that have minimal collinearity by using a continuous projection operation. The algorithm begins by randomly selecting a wavelength from the pool of candidate variables and then calculates its projection onto the remaining wavelengths. The algorithm then compares the sizes of all projection vectors and selects the wavelength with the largest projection vector as the subsequent choice. The process is iterated over time, projecting based on the current set of selected wavelengths in each iteration and incorporating new wavelengths until a predetermined number of wavelengths is reached or a stopping criterion is satisfied.

2.3.4 Regression Model

The NIR analysis establishes a correlation between the spectral characteristics of the sample and its mass parameters, constructs a calibration model, and subsequently utilizes this model to predict the composition and properties of the sample accurately [26]. The quantitative or qualitative relationship between the two can be determined by associating them with mathematical relationships, specifically the regression model [7, 10–12, 25]. After establishing the model, once the NIR spectrum of the unknown sample is measured, the quality parameters of the sample can be predicted based on both the calibration model and its corresponding NIR spectrum. The PLSR method in spectral analysis utilizes the spectral matrix X and concentration matrix Y for principal component analysis, followed by regression analysis using principal factors to establish a model. This approach offers the advantages of simplicity and ease of fitting. The present study applies this method to analyze the spectral characteristic parameters of fresh apricots, aiming to quantitatively predict their inherent quality. The data underwent PLSR analysis using the Unscrambler software. The regression model of PLSR can be expressed as follows:

$$Y = \mathrm{b}X + a \tag{5}$$

where b represents the vector of the regression coefficient and a represents the residual of the model.

The evaluation criteria for the model include the coefficient of determination, root-mean-square error, and Euclidean distance. The closer the Coefficient of determination (R^2) approaches 1, the more accurate the predicted value aligns with the true value. The evaluation of prediction accuracy for the model is based on the root mean square error (RMSEP) of the prediction set. Euclidean Distance (D) represents the shortest distance between two points or vectors in space and effectively reflects sample similarity [17]. The calculation formula is as follows:

$$R^2 = 1 - \frac{\sum_{i=1}^{n} \left(\hat{y_i} - \overline{y}\right)^2}{\sum_{i=1}^{n} (y_i - \overline{y})^2} \tag{6}$$

$$RMSEP = \sqrt{\frac{1}{n}\sum_{i=1}^{n} (y_i - Z_i)} \tag{7}$$

where n represents the number of samples in the prediction set, Yi and Zi represent the measured value and predicted value of the ith sample in the prediction set, respectively, and $\overline{y}$ represents the average measured value of all samples in the prediction set.

$$D = \sqrt{\sum_{k=1}^{\mathrm{n}} \left(A_{ik} - \overline{A_k}\right)^2} \tag{8}$$

where k (k = 1… n) represents the wavelength point, A_{ik} represents the spectrum of the i sample at the k wavelength point collected from the machine, and $\overline{A_k}$ represents the spectral value of the host at the k wavelength point.

3 Results and Discussion

3.1 *Spectrum of Fresh Apricot Samples*

The diffuse reflection spectrum information of fresh apricots was collected using two high-spectral instruments, namely HSI-1 and HSI-2, within the sampling range of 900–1700 nm. The spectra obtained from HSI-1 and HSI-2 for the fresh apricot samples are presented in Fig. 3a and b respectively. In the 900–1700 nm region, spectral data is influenced by a diverse range of chemical bonds and functional groups present in fruits, with C–H, O–H, and N–H bonds being widely prevalent in the

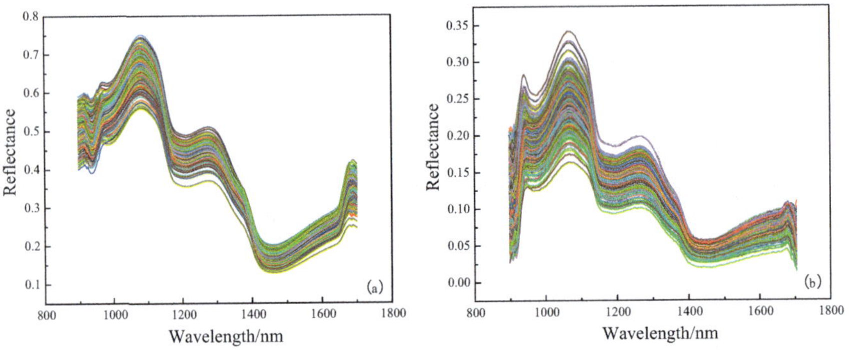

Fig. 3 Primary spectral curve **a** HSI-1 and **b** HSI-2

soluble solids of fruits. The variation trend of all spectral curves remains consistent; however, the reflectance of the sampled spectrum exhibits a certain vertical displacement. This phenomenon may arise from instrument errors induced by optical components and discrepancies in spectral efficiency.

3.2 Regression Model Before Spectral Standardization

To visually compare the spectral data collected by the two instruments, the average spectral curves of fresh apricot samples were separately plotted for each band within the 900–1700 nm range using HSI-1 and HSI-2 instruments, as shown in Fig. 4. The change trend of the two spectral curves is essentially identical; however, there is a significant deviation in the spectral reflectance across the entire band range, with HSI-2 exhibiting significantly lower reflectance compared to HSI-1. The data collected by the two spectrometers exhibit significant disparities, which may lead to divergent predictions when models are trained on these datasets. Such discrepancies have the potential to compromise model accuracy, resulting in inconsistent performance across different datasets. To ensure the universality of the model constructed by HSI-1, it is necessary to calibrate the spectral signals collected by HSI-2.

To validate the predictive performance of spectral modeling obtained from HSI-1, we will develop PLSR models using HSI-1 spectrometer data and apply them to both HSI-1 and HSI-2 spectral datasets. The results of the modeling are shown in Table 1. The predictive performance of this model is commendable for data collected by its own instrument ($R_p^2 = 0.9467$, RMSEP = 0.1541). However, when applied to data collected by HSI-2, the model's R_p^2 decreases to 0.1541 while RMSEP increases to 1.6161 as a result. Consequently, the stability of the model diminishes and significant deviations occur.

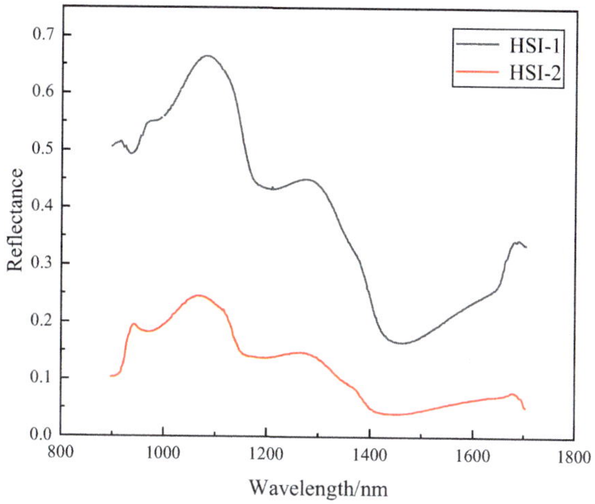

Fig. 4 Average spectrogram of fresh apricot samples

Table 1 The PLS model of fresh apricot SSC was established based on HSI-1

Prediction data set	HSI-1 correction set		Prediction set	
	R_c^2	RMSEC	R_p^2	RMSEP
HSI-1	0.9532	0.1342	0.9467	0.1514
HSI-2			0.1541	1.6161

The prediction results of SSC are shown in Fig. 5 when applying the spectral model established by HSI-1 to the fresh apricot spectral data collected using HSI-1. Furthermore, Fig. 6 displays the prediction results for SSC when utilizing the spectral model developed with HSI-2 on the fresh apricot spectral data obtained from HSI-2. The black line in the figure represents the linear function fitted to the prediction results. The proximity of each point to this line indicates the level of accuracy achieved by the model's predictions. Conversely, points that deviate from the line suggest lower levels of accuracy. The application of the model highlights the disparities in predictions between HSI-1 and HSI-2 due to their specific spectral responses. Therefore, to enhance the robustness of the model across different instruments, a standardized algorithm is employed to correct spectral data and facilitate model transfer between instruments.

3.3 Regression Model After Spectral Normalization

The average spectra of fresh apricot spectral data from HSI-1 and HSI-2, corrected using the DS algorithm, are depicted in Fig. 7. It can be observed from the figure that most bands of the two spectral curves exhibit significant overlap after correction; however, there still exist notable disparities in reflectance among certain bands even

Fig. 5 The data collected by the HSI-1 instrument is applied to the prediction effect of the HSI-1 model

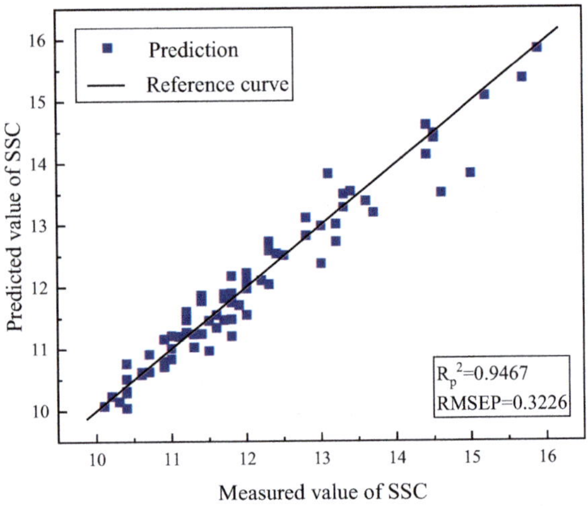

Fig. 6 The data collected by the HSI-2 instrument is applied to the prediction effect of the HSI-1 model

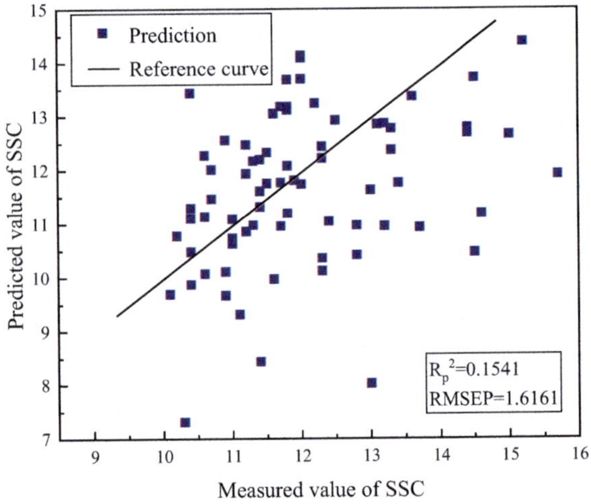

after amplification. The figure displays three bands that exhibit noticeable deviations, specifically at 953–1033 nm, 1045–1135 nm, and 1170–1307 nm. These deviations occur due to significant variations in sensor response, dark current, and noise levels across different bands.

The average spectrum of fresh apricot spectral data collected by HSI-1 and HSI-2, after being corrected using the SLRDS algorithm, is depicted in Fig. 8. Following the SLRDS correction, there is an increased overlap of spectral reflectance bands, with only a minor deviation observed within the range of 966–1010 nm and 1052–1125 nm. The SLRDS algorithm significantly enhances the spectral data similarity between HSI-1 and HSI-2 instruments.

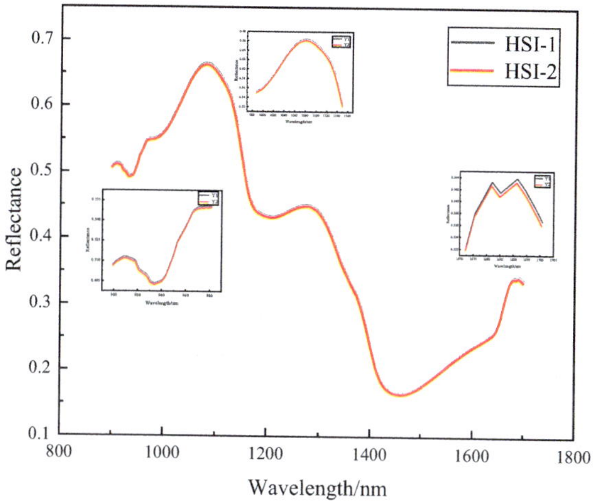

Fig. 7 Average spectrum corrected by DS

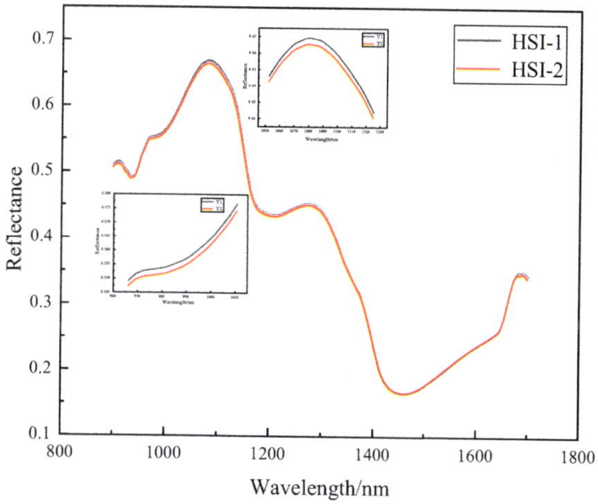

Fig. 8 Average spectrum corrected by SLRDS

The transfer matrix between the spectral data of HSI-1 and HSI-2 calibration sets was established using SBC, DS, and SLRDS algorithms. Subsequently, the spectral data measured on HSI-2 were standardized by applying the transfer matrix. The Euclidean distance was calculated between the standardized spectral data of HSI-2 samples and the spectral data of HSI-1. Then, the established spectral model from HSI-1 was applied to predict the standardized fresh apricot dataset collected by HSI-2. The prediction results are presented in Table 2. After standardization, the Euclidean distance between spectra was significantly reduced, thereby enhancing the transmission effect of the model. Among all tested algorithms, the SLRDS algorithm demonstrated superior performance. When predicting SSC of fresh apricot across different instruments, R_p^2 increased from 0.1541 to 0.9199, and RMSEP decreased from 1.6161 to 0.3935. The application of spectral standardization can effectively

Table 2 The prediction results of the model after spectral standardization

Algorithm	Spectral standardization		Model transfer	
	D_{max}	D_{min}	R_p^2	RMSEP
None	4.2781	3.1440	0.1541	1.6161
SBC	–	–	0.3848	1.1055
DS	0.6540	0.0871	0.8762	0.4885
SLRDS	0.5461	0.0939	0.9199	0.3935

mitigate the impact of instrument variations on SSC determination in hyperspectral fresh apricot analysis, thereby enhancing the robustness and reliability of the model for model transfer across different instruments.

3.4 *Extract Variable*

The spectral data not only contains the information pertaining to the sample but also introduces certain extraneous information, which requires a significant amount of time to achieve optimal detection efficacy. The model optimization process involves standardizing the spectrum using SLRDS and then utilizing the SPA algorithm to extract relevant feature wavelengths, eliminate irrelevant information, and remove bands with significant spectral deviation. This approach effectively achieves a reduction in data dimensionality. The root-mean-square error diagram and the optimal distribution diagram of characteristic parameters obtained from the spectral data of HSI-1 after applying the SPA algorithm are depicted in Figs. 9 and 10, respectively. The figure illustrates that the model achieves the minimum root-mean-square error value when there are 13 variables in the final number. The 13 characteristic wavelengths include 907, 912, 926, 973, 1176, 1317, 1317, 1374, 1402, 1463, 1539, 1628 and 1690 nm.

The root-mean-square error diagram and the optimal distribution diagram of characteristic parameters obtained from the spectral data of HSI-2 using the SPA algorithm are depicted in Figs. 11 and 12, respectively. The following 13 characteristic wavelengths were selected: 907, 926, 954, 1015, 1147, 1176, 1355, 1402, 1454, 1529, 1628, 1680 and 1690 nm.

The characteristic wavelengths obtained by the two instruments are 907 nm, 926 nm, 1176 nm, 1402 nm, and 1628 nm respectively, while the remaining wavelengths exhibit variations. The 13 characteristic wavelengths obtained from HSI-2 were consolidated into a single wavelength, while the characteristic wavelengths extracted by both instruments were combined to form a composite wavelength. The optimization results of the model, based on the full wavelength, single wavelength, and combined wavelengths after transmission, are presented in Table 3. The table reveals that, compared to the full wavelength, both single and combined wavelengths show a decrease in the number of wavelength variables. Additionally, there is a

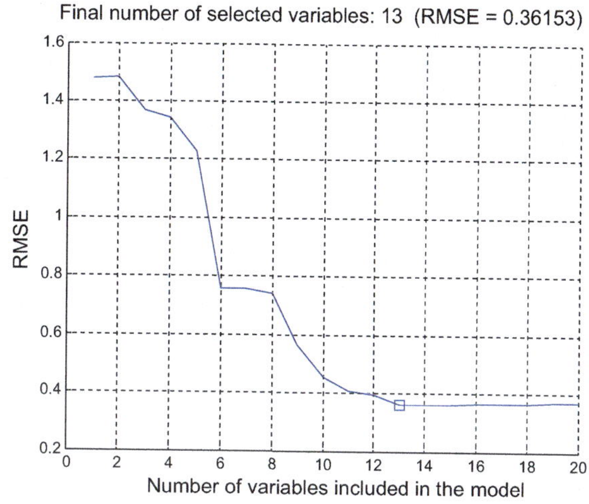

Fig. 9 Relationship between the number of characteristic parameters and root mean square error

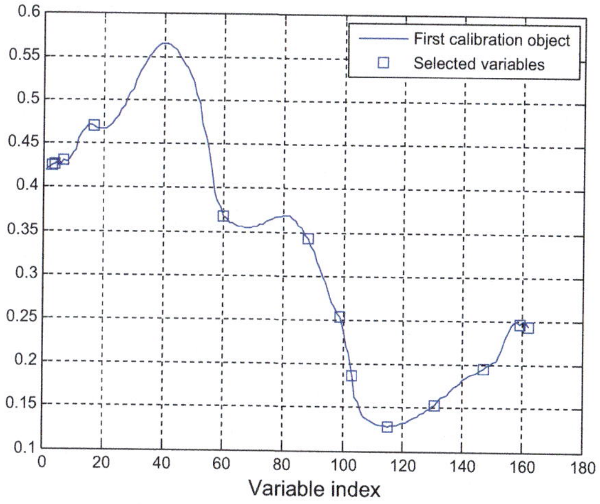

Fig. 10 Preferred distribution of characteristic parameters

reduction in the European distance of spectral data, an increase in standardization degree, and an improvement in model prediction performance. Specifically, R_p^2 increases from 0.9199 to 0.9405 and 0.9244 respectively, while RMSEP decreases from 0.3935 to 0.3409 and 0.3818. The reasonable combination of the spectral standardization method and feature band selection method can effectively enhance the model's transmission capability and prediction accuracy, ultimately leading to improved performance.

The prediction results of the PLSR model, established using full wavelength, single wavelength, and combined wavelength data from fresh apricot soluble solid content (SSC), are presented in Figs. 13, 14, and 15 respectively.

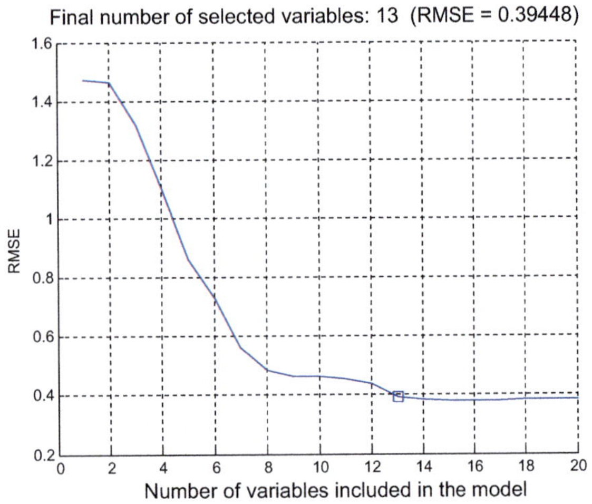

Fig. 11 Relationship between the number of characteristic parameters and root mean square error

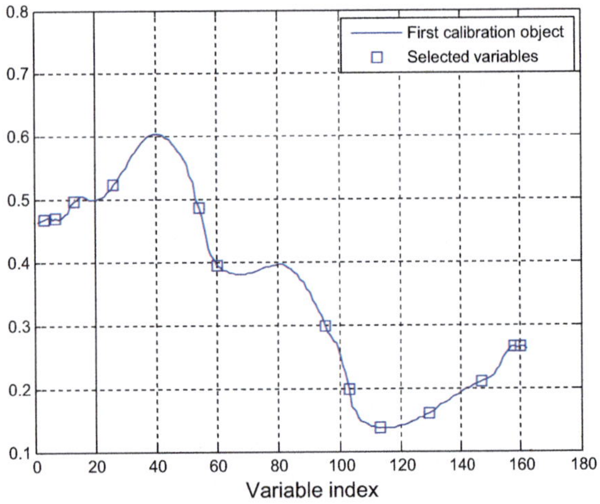

Fig. 12 Preferred distribution of characteristic parameters

Table 3 The prediction results of the model after selecting the variables

Wavelength	Characteristic variable/full variable	Spectral standardization		Model transfer			
		Dmax	Dmin	R_c^2	RMSEC	R_p^2	RMSEP
Full wavelength	162/162	0.5461	0.0939	0.9279	0.3945	0.9199	0.3935
Single wavelength	13/162	0.1507	0.0291	0.9343	0.3765	0.9405	0.3409
Combined wavelength	21/162	0.1686	0.0265	0.9262	0.3993	0.9244	0.3818

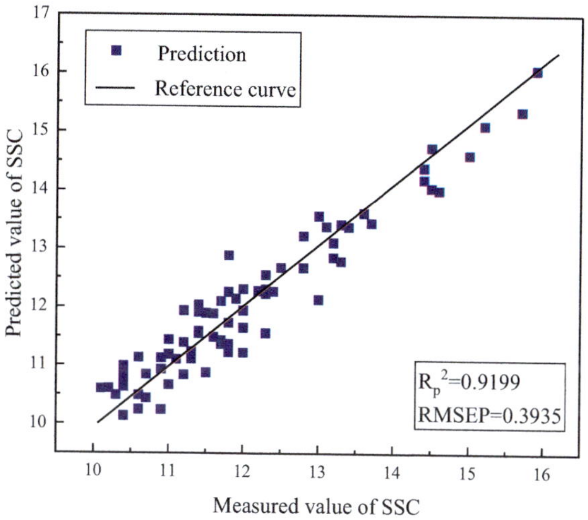

Fig. 13 Prediction results of SSC in fresh apricot based on full wavelength

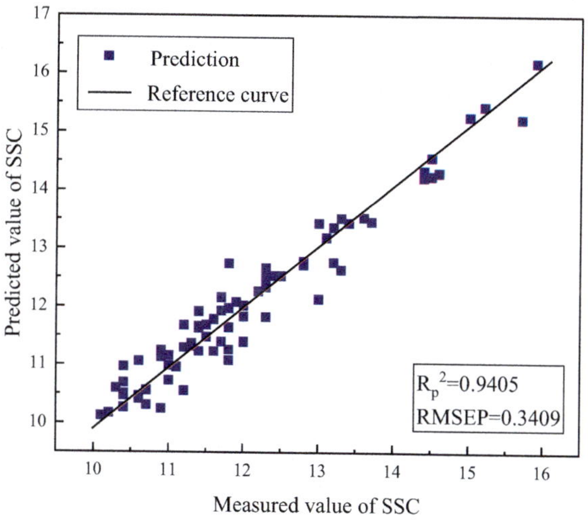

Fig. 14 Prediction results of SSC in fresh apricot based on single wavelength

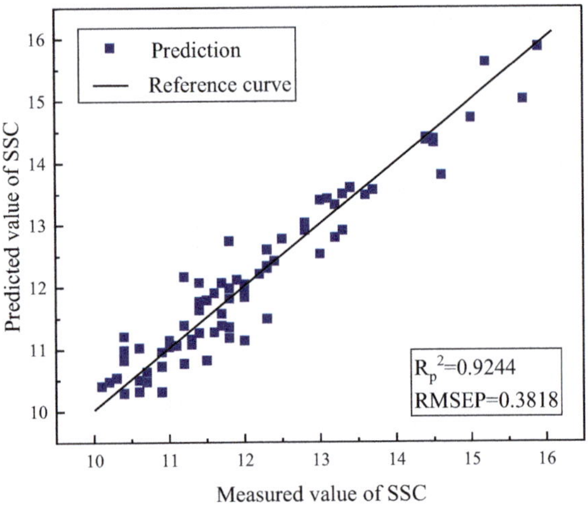

Fig. 15 Prediction results of SSC in fresh apricot based on combined wavelength

4 Conclusion

In this study, fresh apricots were selected as the research subject, and HSI-1 and HSI-2 spectrometers were utilized to collect spectral information from the samples, respectively. The impact of different spectral correction algorithms and characteristic wavelengths on the transmission of SSC detection model between the two instruments was investigated in order to address technical issues in detection and enhance efficiency and accuracy.

The established PLSR model using full-band spectral information for HSI-1 demonstrated excellent performance in predicting the collected spectra, with an R_p^2 value of 0.9467 and an RMSEP value of 0.3226. However, when utilizing the spectrum for HSI-2 acquisition, R_p^2 is equal to 0.1541 and RMSEP is 1.6161, which renders the model invalid. The spectral data collected by HSI-2 were corrected using three methods, namely SBC, DS, and SLRDS, in order to achieve accurate model transmission. After comparison, the SLRDS model demonstrates superior transmission effectiveness. Following the standardization of the spectrum collected by HSI-2 using the SLRDS algorithm, model R_p^2 significantly increases from 0.1541 to 0.9307, while RMSEP decreases substantially from 1.6161 to 0.3675.

After standardizing the spectral data, SPA was used to extract the characteristic wavelengths of HSI-1 and HSI-2 spectrometers, resulting in the selection of 13 individual wavelengths and 21 combined wavelengths. The optimal modeling effect was achieved by selecting a single wavelength after conducting model transfer, resulting in a reduction of variables to only 8.02% of the full wavelength range. When applied for HSI-2 spectrum prediction, R_p^2 reached 0.9405 and RMSEP was measured at 0.3409, thereby establishing a foundation for model transfer across different instruments.

The amalgamation of spectral standardization algorithms with characteristic wavelengths confers advantages during model transmission enhancement while effectively addressing issues related to inter-instrument model sharing limitations and low stability across varying conditions. His study offers valuable insights into facilitating cross-instrumental sharing of spectral models by providing a methodological framework as reference.

Author Contribution Runrun Wang: Data curation; writing; experiment; Shujuan Zhang: Writing-review; formulation or evolution of overarching research goals and aims; Zhao Zhang: Writing-review.

Conflict of Interest The authors declare no conflicts of interest.

References

1. Chemometrics et al (2017) Chemometrics in spectroscopy bias and slope correction. Chem Chem 3036
2. Chen Y, Wang Z (2019) Cross components calibration transfer of NIR spectroscopy model through PCA and weighted ELM-based TrAdaBoost algorithm. Chemometr Intell Lab Syst 192:103824–103824
3. Chun WS, Song JD, Lee HK et al (2024) Deep learning algorithm development for early detection of *Botrytis cinerea* infected strawberry fruit using hyperspectral fluorescence imaging. Postharv Biol Technol 214:112918
4. Ding G, Jin K, Chen X et al (2024) Non-destructive prediction of ready-to-eat kiwifruit firmness based on Fourier transform near-infrared spectroscopy. Postharv Biol Technol 212:112908
5. Dong X, Dong J, Li Y et al (2019) Maintaining the predictive abilities of egg freshness models on new variety based on VIS-NIR spectroscopy technique. Comput Electron Agric 156:669–676
6. Fan S, Li J, Xia Y et al (2019) Long-term evaluation of soluble solids content of apples with biological variability by using near-infrared spectroscopy and calibration transfer method. Postharv Biol Technol 151:79–87
7. Gao R, Li J, Dong L et al (2024) Accurate analysis of coal calorific value using NIRS-XRF: utilizing RF classification and PLSR subtype modeling. Microchem J 131:709
8. Grelet C, Pierna FAJ, Dardenne P et al (2017) Standardization of milk mid-infrared spectrometers for the transfer and use of multiple models. J Dairy Sci 100(10):7910–7921
9. Huang ZX, Sanaeifar A, Tian Y et al (2021) Improved generalization of spectral models associated with Vis-NIR spectroscopy for determining the moisture content of different tea leaves. J Food Eng 293:110374
10. Li J, Ma Z, Dai H et al (2024) Application of PLSR in correlating sensory and chemical properties of middle flue-cured tobacco leaves with honey-sweet and burnt flavour. Heliyon 10(8):e29547
11. Li J, Chu X (2024) Dense interpolation—wavelength shifting—mean spectra subtraction correction: a new calibration transfer algorithm for near-infrared spectroscopy instruments on crude oil fast assay. Fuel 370:131820
12. Li Y, Sun H, Zheng Y et al (2024) Combined gramian angular difference field image coding and improved mobile vision transformer for determination of apple soluble solids content by Vis-NIR spectroscopy. J Food Compos Anal 131:106200
13. Luiza ABB, Vivianny ASP, Danielle KMMT et al (2017) Calibration transfer of flour NIR spectra between benchtop and portable instruments. Anal Methods 9(21):3184–3190

14. Mitrofanova IV, Mitrofanova OV, Lesnikova-Sedoshenko NP et al (2020) Some features of obtaining new breeding forms of apricot in vitro. Acta Hortic 1290
15. Pang L, Wang J, Men S et al (2021) Hyperspectral imaging coupled with multivariate methods for seed vitality estimation and forecast for *Quercus variabilis*. Spectrochim Acta Part A Mol Biomol Spectrosc 245:118888–118888
16. Rahmia DN et al (2022) Non-destructive measurement of malic acid content of sapodilla fruit using visible near infrared (VisNIR) spectroscopy with variations in storage temperature. IOP Conf Ser Earth Environ Sci 1116(1)
17. Song XP, Wu SH, Li G et al (2022) A calibration set selection method fusing multi-component based on Euclidean distance in spectral analysis of complex solution. Infrared Phys Technol 123
18. Sonia S, Ashima A, Himanshu S et al (2023) An overview of dietary flavonoids as a nutraceutical nanoformulation approach to life-threatening diseases. Curr Pharm Biotechnol 24(14):1740–1773
19. Sun XT, Zhang P, Shen ZW et al (2021) Investigation on spectral standardization among multi-channel of an on-line near-infrared spectrometer. Vib Spectrosc 113
20. Wang HH, Yuan H, Xiong ZX (2024) Transfer of near infrared calibration for gasoline octane number based on screening consistent wavelengths combined with direct standardization algorithm. J Near Infrared Spectrosc 32(1–2):38–45
21. Wang HH, Li PH, Xiong ZX et al (2024) Pulpwood holocellulose model transfer based on screening with consistent and stable signals algorithm. Anal Sci Int J Japan Soc Anal Chem
22. Xia K, Wu T, Zhang S et al (2024) A new method for high-precision estimation of soil organic matter using two-dimensional correlation spectroscopy—to support collaborative use of global open soil spectral libraries. Geoderma 445:116877
23. Xu LJ, Chen YJ, Wang XH et al (2023) Non-destructive detection of kiwifruit soluble solid content based on hyperspectral and fluorescence spectral imaging#13. Front Plant Sci 13:1075929
24. Yan P, Li G, Wang W et al (2024) Qualitative and quantitative detection of microplastics in soil based on LIF technology combined with OOA-ELM/SPA-PLS. Microchem J 201:110632
25. Yang J, Kong L, Ye H (2024) Surface hardness determination of laser cladding using laser-induced breakdown spectroscopy and machine learning (PLSR, CNN, ResNet, and DRSN). Appl Opt 63(10):2509–2517
26. Yang L, Meng L, Gao H et al (2021) Building a stable and accurate model for heavy metal detection in mulberry leaves based on a proposed analysis framework and laser-induced breakdown spectroscopy. Food Chem 338:127886–127886
27. Yu X, Lu H, Wu D (2018) Development of deep learning method for predicting firmness and soluble solid content of postharvest Korla fragrant pear using Vis/NIR hyperspectral reflectance imaging. Postharv Biol Technol 141:39–49

The manufacturer's authorised representative in the EU is Springer Nature Customer Service Centre GmbH, Europaplatz 3, 69115 Heidelberg, Germany. If you have any concerns regarding our products, please contact ProductSafety@springernature.com

Printed and bound by CPI Group (UK) Ltd, Croydon, CR0 4YY
01/10/2026
02225919-0010